OXFORD STATISTICAL SCIENCE SERIES

OXFORD STATISTICAL SCIENCE SERIES

A. C. Atkinson: *Plots, transformations, and regression*
M. Stone: *Coordinate-free multivariable statistics*
W. J. Krzanowski: *Principles of multivariate analysis: a user's perspective*
M. Aitkin, D. Anderson, B. Francis, and J. Hinde: *Statistical modelling in GLIM*
Peter J. Diggle: *Time series: a biostatistical introduction*
Howell Tong: *Non-linear time series: a dynamical system approach*
V. P. Godambe: *Estimating functions*
A. C. Atkinson and A. N. Donev: *Optimum experimental designs*
U. N. Bhat and I. V. Basawa: *Queuing and related models*
J. K. Lindsey: *Models for repeated measurements*
N. T. Longford: *Random coefficient models*
P. J. Brown: *Measurement, regression, and calibration*
Peter J. Diggle, Kung-Yee Liang, and Scott L. Zeger: *Analysis of longitudinal data*
J. I. Ansell and M. J. Phillips: *Practical methods for reliability data analysis*
J. K. Lindsey: *Modelling frequency and count data*
J. L. Jensen: *Saddlepoint approximations*
Steffen L. Lauritzen: *Graphical models*
A. W. Bowman and A. Azzalini: *Applied smoothing techniques for data analysis*

Applied Smoothing Techniques for Data Analysis

The Kernel Approach with S-Plus Illustrations

ADRIAN W. BOWMAN

Department of Statistics
University of Glasgow

and

ADELCHI AZZALINI

Department of Statistical Sciences
University of Padova, Italy

CLARENDON PRESS • OXFORD

This book has been printed digitally and produced in a standard specification in order to ensure its continuing availability

OXFORD
UNIVERSITY PRESS

Great Clarendon Street, Oxford OX2 6DP
Oxford University Press is a department of the University of Oxford.
It furthers the University's objective of excellence in research, scholarship, and education by publishing worldwide in

Oxford New York

Auckland Cape Town Dar es Salaam Hong Kong Karachi
Kuala Lumpur Madrid Melbourne Mexico City Nairobi
New Delhi Shanghai Taipei Toronto
With offices in
Argentina Austria Brazil Chile Czech Republic France Greece
Guatemala Hungary Italy Japan South Korea Poland Portugal
Singapore Switzerland Thailand Turkey Ukraine Vietnam

Oxford is a registered trade mark of Oxford University Press
in the UK and in certain other countries

Published in the United States
by Oxford University Press Inc., New York

Oxford is a registered trade mark of Oxford University Press
in the UK and in certain other countries

Published in the United States
by Oxford University Press Inc., New York

Reprinted 2004

ISBN 0-19-852396-3

To our families

PREFACE

This book is about the use of nonparametric smoothing tools in practical data analysis problems. There already exist a number of books in this general area, and these give excellent descriptions of the theoretical ideas, and include good illustrations of applications. The aim of this text is to complement the existing books in two ways. Firstly we have focused particularly on intuitive presentation of the underlying ideas across a wide variety of different contexts and data structures. The level of mathematical treatment is correspondingly rather light and where details are appropriate these have been separated from the main text. The emphasis is on introducing the principles of smoothing in an accessible manner to a wide audience. Secondly we have emphasised practical issues of inference, where well defined procedures for this exist, rather than of estimation. These aims give particular prominence to the role of graphics, firstly to communicate basic ideas in an informal, intuitive way, and secondly to provide tools for exploring and analysing a variety of different types of data.

This monograph aims to introduce nonparametric smoothing to statisticians and researchers in other scientific areas who seek a practical introduction to the topic. In view of its style of presentation, the text could also form the basis of a course in nonparametric smoothing suitable for students of statistics. Both density estimation and nonparametric regression are covered although there is considerably more emphasis on the latter in view of its more numerous applications. For each of these two areas a general introduction is provided by one chapter focusing on basic ideas and graphical explorations of data and another which prepares the ground for inferential techniques. The remainder of the text extends these ideas to a variety of different data structures. The final chapter provides a brief introduction to the more general tools provided by generalised additive models. Throughout the book we have endeavoured to provide a wide variety of illustrations and exercises.

In the last few years, the development of a number of different areas of statistics has been greatly influenced by ever more powerful computing environments. The area of nonparametric smoothing is a particular example of this, where techniques can now be implemented in a practical manner, in computing environments which are widely available. At the end of each section of this book, examples are given of how graphics and analysis can be carried out in the S-Plus computing environment. Software specifically designed to support the illustrations discussed in the book is freely available. It should be emphasised that the book can easily be used without reference to this. However, for those who have access to S-Plus, the software will provide a means of exploring the concepts discussed in the text, and to provide simple tools for analysis which can be applied in a straightforward and accessible manner. Novice users will be able to explore data without the need to learn the detailed syntax of S-Plus, while experienced users will be able to extend these tools and analyses. Details of how to obtain the S-Plus material, which includes functions, data and scripts, are given in an appendix.

Over the years, our own work has benefited greatly from the support and advice of a large number of friends and colleagues with interest in ideas of non-parametric smoothing. There are far too many names to list here, but special mention is due to Mike Titterington, who played a major role in introducing one of us to this research area, and to Chris Jones, who has often been a source of helpful advice and who was kind enough to make comments on this book before publication. Some of the initial work for the book was carried out in preparation for a short course given in conjunction with a conference of the Royal Statistical Society, held in Newcastle, in September 1994. We are very grateful for the opportunity to present this course, and for the support given by the RSS in general, and Peter Diggle and Robin Henderson in particular. Further development took place while one of us (AWB) enjoyed the hospitality of the Department of Statistics in the Australian National University.

Important contributions to the ideas involved in the book, to their presentation, and also to their S-Plus implementation, have been made by a succession of students. This includes Eileen Wright, Stuart Young, Angela Diblasi, Mitchum Bock and Adrian Hines. A large number of colleagues were kind enough to grant access to previously unpublished data, to advise on their context and in some cases to make significant contributions to the analysis reported in the book. This includes Niall Anderson, Dave Borchers, Steve Buckland, Charis Burridge, David Crompton, Peter Diggle, Andy Dugmore, Malcolm Faddy, Tony Gatrell, Roger Gosden, Julia Kelsall, Monty Priede, Paolo Saviotti, Marian Scott and Peng Weidong. Finally, but by no means least, we are grateful to a number of people whose knowledge of S-Plus and LaTeX 2_ε greatly exceeded our own and who were willing to allow us to reap the benefits of their expertise. This includes Monica Chiogna, Michal Jaegerman and Brian Ripley.

CONTENTS

1

DENSITY ESTIMATION FOR EXPLORING DATA

1.1 Introduction

The concept of a probability density function is a central idea in statistics. Its role in statistical modelling is to encapsulate the pattern of random variation in the data which is not explained by the other structural terms of a model. In many settings this role, while important, is a secondary one, with the principal focus resting on the nature of covariate or other effects. However, there are also situations where the detailed shape of the underlying density function is itself of primary interest. In this chapter, a number of examples of this are given and the ideas behind the construction of a smooth estimate of a density function are introduced. Some of the main issues associated with using these estimates are raised.

1.2 Basic ideas

In a study of the development of aircraft technology, Saviotti and Bowman (1984) analysed data on aircraft designs. The first author subsequently collected more extensive data, mainly from Jane's (1978), on six simple characteristics of aircraft designs which have appeared during the twentieth century. These six characteristics are:

- total engine power (kW);
- wing span (m);
- length (m);
- maximum take-off weight (kg);
- maximum speed (km h^{-1});
- range (km).

The aim of this study was to explore techniques for describing the development of this technology over time, and in particular to highlight particular directions in which this development occurred. Techniques which successfully described the patterns in this well understood area might then be applied to good effect in other, less well known, areas.

Clearly, events such as two world wars have had an enormous impact on aircraft development. In view of this, the data will be considered in separate subgroups corresponding to the years 1914–1935, 1936–1955 and 1956–1984. The left panel of Fig. 1.1 displays a histogram of the data on wing span from the third time period. Since all of the six variables have markedly skewed distributions, each will be examined on a log scale.

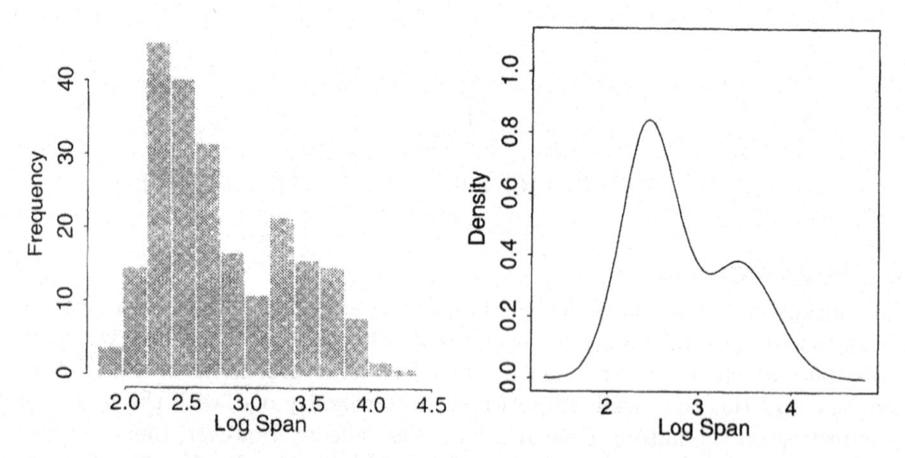

FIG. 1.1. Histogram and density estimate of the aircraft span data, on a log scale, for the third time period.

The histogram is, of course, a widely used tool for displaying the distributional shape of a set of data. More specifically, its usefulness lies in the fact that it indicates the shape of the underlying density function. For example, with the span data it is clear that some skewness exists, even on the log scale.

The right panel of Fig. 1.1 displays an alternative estimate of the density function as a smooth curve. In order to discuss the construction of estimators of this type, it is helpful to consider first the construction of a histogram. This begins by dividing the sample space into a number of intervals. Each observation contributes a 'box' which is then placed over the appropriate interval. This is illustrated in the left panel of Fig. 1.2, which uses a small subsample of the span data for the purpose of illustration. If y denotes the point at which the density $f(y)$ must be estimated, then the histogram may be written as

$$\tilde{f}(y) = \sum_{i=1}^{n} I(y - \tilde{y}_i; h),$$

where $\{y_1, \ldots, y_n\}$ denote the observed data, $\tilde{y}_i$ denotes the centre of the interval in which y_i falls and $I(z; h)$ is the indicator function of the interval $[-h, h]$. Notice that further scaling would be required to ensure that $\tilde{f}$ integrates to 1.

Viewed as a density estimate, the histogram may be criticised in three ways.

- Information has been thrown away in replacing y_i by the central point of the interval in which it falls.
- In most circumstances, the underlying density function is assumed to be smooth, but the estimator is not smooth, due to the sharp edges of the boxes from which it is built.

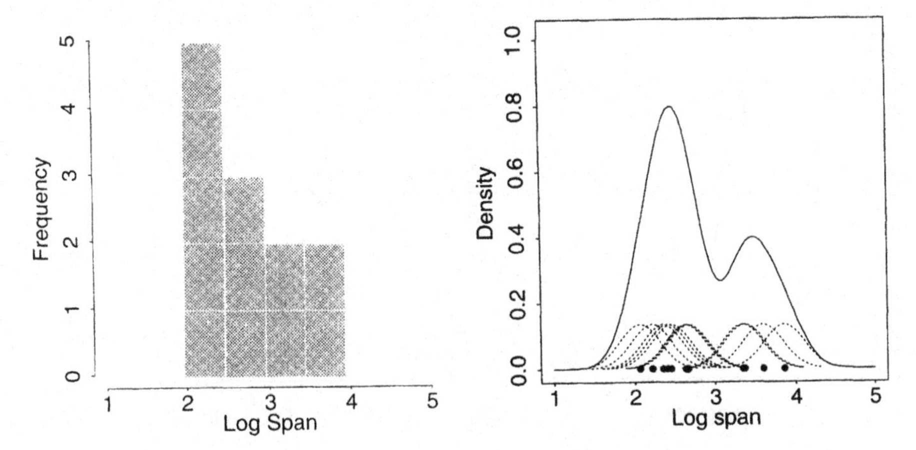

FIG. 1.2. Illustration of the construction of a histogram and density estimate from a subsample of the aircraft span data on a log scale.

◇ The behaviour of the estimator is dependent on the choice of width of the intervals (or equivalently boxes) used, and also to some extent on the starting position of the grid of intervals.

Rosenblatt (1956), Whittle (1958) and Parzen (1962) developed an approach to the problem which removes the first two of these difficulties. First, a smooth *kernel* function rather than a box is used as the basic building block. Second, these smooth functions are centred directly over each observation. This is illustrated in the right panel of Fig. 1.2. The kernel estimator is then of the form

$$\hat{f}(y) = \frac{1}{n} \sum_{i=1}^{n} w(y - y_i; h), \tag{1.1}$$

where w is itself a probability density, called in this context a kernel function, whose variance is controlled by the parameter h.

It is natural to adopt a function w which is symmetric with mean 0, but beyond that it is generally agreed that the exact shape is not too important. It is often convenient to use for w a normal density function, so that

$$w(y - y_i; h) = \phi(y - y_i; h),$$

where $\phi(z; h)$ denotes the normal density function in z with mean 0 and standard deviation h. Because of its role in determining the manner in which the probability associated with each observation is spread over the surrounding sample space, h is called the *smoothing parameter* or *bandwidth*. Since properties of w are inherited by $\hat{f}$, choosing w to be smooth will produce a density estimate which is also smooth.

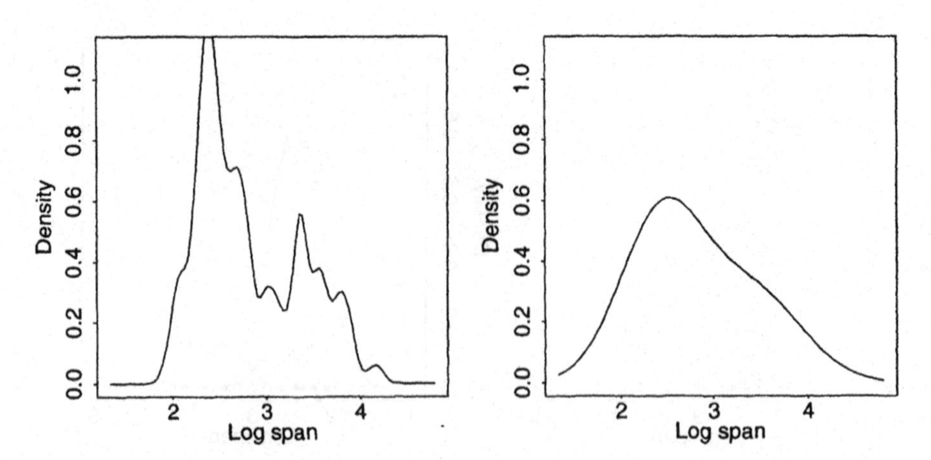

FIG. 1.3. The effect of changing the smoothing parameter on a density estimate of the aircraft span data.

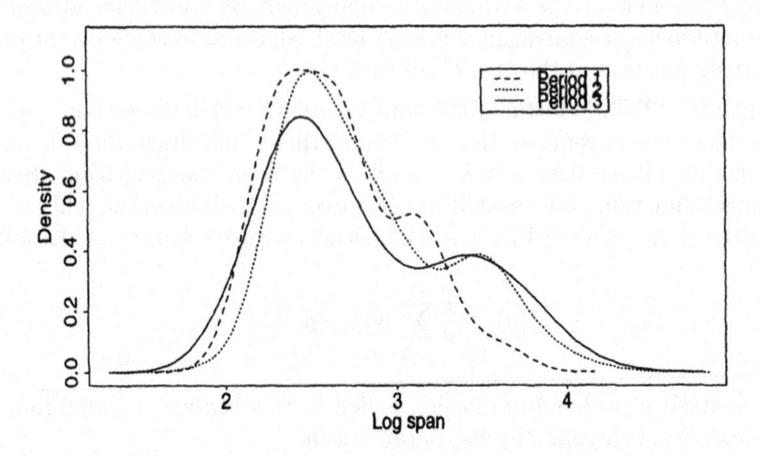

FIG. 1.4. A comparison of density estimates based on the log span data for the time periods 1914–1935, 1936–1955 and 1956–1984.

The third criticism of the histogram still applies to the smooth density estimate, namely that its behaviour is affected by the choice of the width of the kernel function. This is illustrated in Fig. 1.3. When h is small the estimate displays the variation associated with individual observations rather than the underlying structure of the whole sample. When h is large this structure is obscured by smoothing the data over too large a region. Strategies for choosing the smoothing parameter will be discussed in Chapter 2.

One advantage of a smooth density estimate is that comparisons among different groups become easier. For example, Fig. 1.4 displays estimates constructed from the three groups of data corresponding to the time periods mentioned above. It is difficult to superimpose histograms, but density estimates can be plotted together easily and the underlying shapes contrasted more effectively. In this case, the principal modes of the distributions are seen to occur at virtually identical positions, indicating that the most commonly used wing spans have changed very little throughout the century. The third time period displays a larger proportion of the distribution around a subsidiary mode at higher wing spans, as well as an increased proportion in the lower tail.

S-Plus Illustration 1.1. A density estimate of the log span data

Figure 1.1 was constructed with the following S-Plus *code. The* `provide.data` *and* `sm.density` *functions are supplied in the* `sm` *library which has been written in conjunction with this book. Details on obtaining and using this library are given in an Appendix.*

```
provide.data(aircraft)
y <- log(Span[Period==3])
par(mfrow=c(1,2))
hist(y, xlab="Log Span", ylab="Frequency")
sm.density(y, xlab="Log Span")
par(mfrow=c(1,1))
```

S-Plus Illustration 1.2. Changing the smoothing parameter

Figure 1.3 was constructed with the following S-Plus *code. The parameter* `hmult` *adjusts the default smoothing parameter by multiplying it by the stated value.*

```
provide.data(aircraft)
y <- log(Span[Period==3])
par(mfrow=c(1,2))
sm.density(y, hmult = 1/3, xlab="Log span")
sm.density(y, hmult = 2,   xlab="Log span")
par(mfrow=c(1,1))
```

An interactive exploration of the effect of changing the smoothing parameter can be obtained by adding the argument `panel=T` *to the* `sm.density` *function. This will launch a mouse-activated menu which also allows the possibility of an animated display.*

S-Plus Illustration 1.3. Comparing density estimates

Figure 1.4 was constructed with the following S-Plus *code.*

```
provide.data(aircraft)
y1 <- log(Span)[Period==1]
y2 <- log(Span)[Period==2]
```

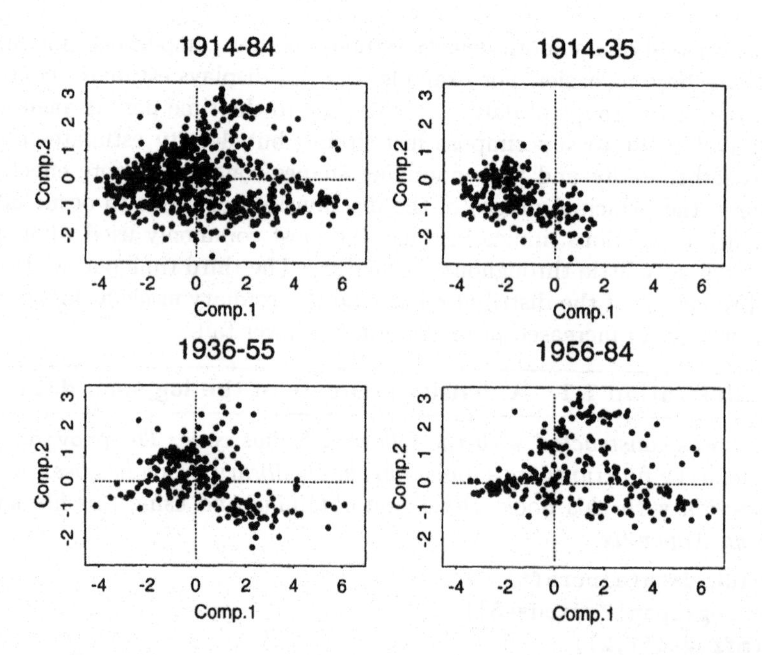

FIG. 1.5. Scatterplots of the first two principal component scores for the aircraft data. The first plot shows all the data. The remainder display the data for the time periods 1914–1935, 1936–1955 and 1956–1984.

```
y3 <- log(Span)[Period==3]
sm.density(y3, xlab="Log span")
sm.density(y2, add=T, lty=2)
sm.density(y1, add=T, lty=3)
legend(3.5, 1, c("Period 1", "Period 2", "Period 3"), lty=3:1)
```

1.3 Density estimation in two dimensions

The kernel method extends to the estimation of a density function in more than one dimension. The aircraft data are six-dimensional and Fig. 1.5 displays a plot of the first two principal components. The first component can broadly be identified with the 'size' of the aircraft as it is constructed from a mixture of all variables except speed. The second component can broadly be identified from its coefficients as 'speed adjusted for size'.

As a descriptive exercise, a two-dimensional density estimate can be constructed for these data by applying (1.1) with a two-dimensional kernel function in the form

$$\hat{f}(y_1, y_2) = \frac{1}{n}\sum_{i=1}^{n} w(y_1 - y_{1i}; h_1)\, w(y_1 - y_{2i}; h_2),$$

where $\{y_{1i}, y_{2i}; i = 1, \dots n\}$ denote the data and (h_1, h_2) denote the joint smoothing parameters. It would be possible to use a bivariate kernel whose components are correlated but it is convenient, and usually sufficient, to employ a product of univariate components. Figure 1.6 shows the result of this with the data from the third time period, using normal kernel functions. The perspective plot shows that there are in fact three separate modes in the estimate, a feature which is not immediately clear from the simple scatterplot. The effect of the smoothing parameter is displayed in the remaining two panels, where a small value produces spurious peaks and troughs, while a large value obscures the individual modes. As in the one-dimensional case, the choice of this parameter can be important. In addition, it would clearly be very helpful to have some means of assessing which features are genuine and which are spurious when using these smoothing techniques.

Figure 1.7 illustrates alternative forms of display for a two-dimensional density estimate. In the left panel the height of the estimate is indicated by grey shading in an 'imageplot'. In the right panel contours are drawn. A standard approach would be to draw these contours at equally spaced heights. However, the contours here have been carefully selected in order to contain specific proportions of the observations. The contour labelled '75' contains the 75% of the observations corresponding to the greatest density heights, and similarly for the contours labelled '50' and '25'. These contours can be chosen easily by evaluating the density estimate at each observation, ranking the density heights and locating the median and quartiles. In this way the display has a construction which is reminiscent of a boxplot, although it has the additional ability to display multimodality through the disjoint nature of some of the contours. Bowman and Foster (1993) discuss this type of display in greater detail. The term 'sliceplot' will be used to differentiate the display from a standard contour plot. With the aircraft data the contours draw clear attention to the multimodal nature of the data. This can be interpreted as a degree of specialisation, with the appearance of aircraft which are fast but not large, and large but not fast.

The main aim of the analysis of the aircraft data was to identify changes over time. An effective way to address this is to plot and compare density estimates for the three separate time periods identified above. This is done in Fig. 1.8. The shape of each density estimate is characterised in a single contour containing 75% of the data and these are superimposed to show clearly the manner in which size and speed have changed in the aircraft designs produced over the century.

S-Plus Illustration 1.4. Density estimates from the aircraft span data

---

Figure 1.6 was constructed with the following S-Plus *code. The* `cex` *and* `zlim` *parameters are used simply to produce more attractive axis scaling.*

```
provide.data(airpc)
pc3 <- cbind(Comp.1, Comp.2)[Period==3,]
par(mfrow=c(2,2))
```

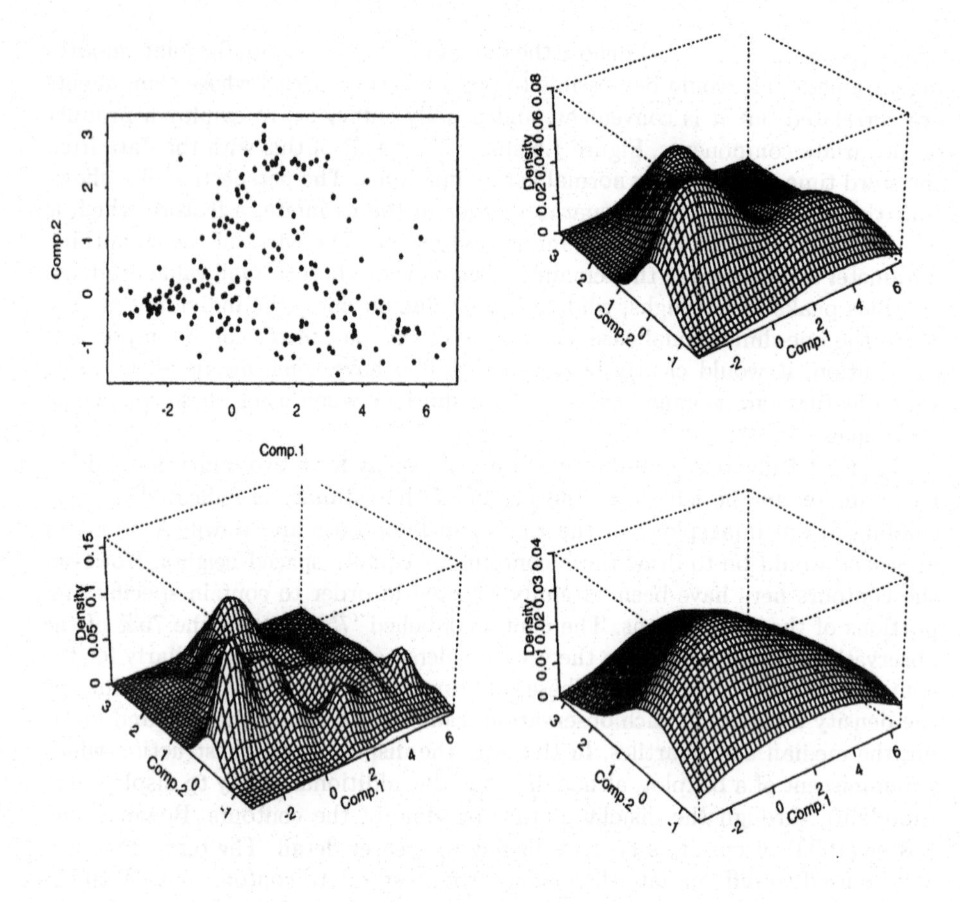

FIG. 1.6. A scatterplot and density estimates based on the log span data for the time period 1956–1984.

```
par(cex=0.6)
plot(pc3)
sm.density(pc3,              zlim=c(0,0.08))
sm.density(pc3, hmult=1/2, zlim=c(0,0.15))
sm.density(pc3, hmult=2,    zlim=c(0,0.04))
par(cex=1)
par(mfrow=c(1,1))
```

S-Plus Illustration 1.5. An imageplot and sliceplot from the aircraft span data

Figure 1.7 was constructed with the following S-Plus *code.*

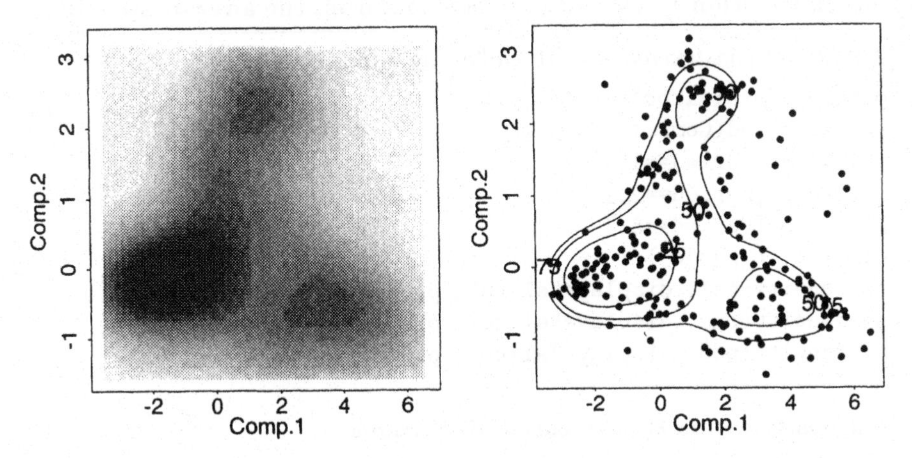

FIG. 1.7. An 'imageplot' and 'sliceplot' of density estimates based on the log span data for the time period 1956–1984.

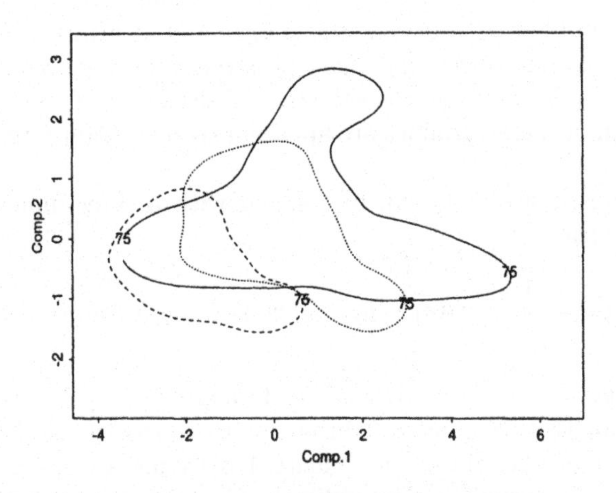

FIG. 1.8. Density estimates, represented as contour plots, based on the log span data for the time periods 1914–1935 (dashed line), 1936–1955 (dotted line) and 1956–1984 (full line).

```
provide.data(airpc)
pc3 <- cbind(Comp.1, Comp.2)[Period==3,]
par(mfrow=c(1,2))
sm.density(pc3, display="image")
sm.density(pc3, display="slice")
par(mfrow=c(1,1))
```

S-Plus Illustration 1.6. Multiple sliceplots from the aircraft span data

Figure 1.8 was constructed with the following S-Plus *code.*

```
provide.data(airpc)
pc  <- cbind(Comp.1, Comp.2)
pc1 <- pc[Period==1,]
pc2 <- pc[Period==2,]
pc3 <- pc[Period==3,]
plot(pc, type="n")
sm.density(pc1, display="slice", props=75, add=T, lty=3)
sm.density(pc2, display="slice", props=75, add=T, lty=2)
sm.density(pc3, display="slice", props=75, add=T, lty=1)
```

1.4 Density estimation in three dimensions

The Old Faithful geyser in Yellowstone National Park exhibits an unusual structure in its pattern of eruption times, and in the length of the waiting times between successive eruptions. Weisberg (1985) collected data on this and proposed a regression model to predict the waiting time. Azzalini and Bowman (1990) described similar data in time series form and identified the relationships among the three variables *waiting time*, *duration* and *subsequent waiting time* to be important in determining the structure of the series. Three clusters in the joint distribution of these variables are apparent by deduction from the marginal scatterplots.

The density estimate (1.1) can by applied with a three-dimensional kernel function in the form

$$\hat{f}(y_1, y_2, y_3) = \frac{1}{n}\sum_{i=1}^{n} w(y_1 - y_{1i}; h_1)\, w(y_2 - y_{2i}; h_2)\, w(y_3 - y_{3i}; h_3),$$

where $\{y_{1i}, y_{2i}, y_{3i}; i = 1, \ldots, n\}$ denote the data and (h_1, h_2, h_3) denote the joint smoothing parameters. It is again convenient to construct the kernel function from the product of univariate components. Displaying a density estimate as a function of three dimensions is more difficult. Contours were used to good effect with two-dimensional data, where a contour is a closed curve, or a set of closed curves if multimodality is present. A contour of a function defined in terms of three arguments is a more unusual object. In fact, it is a closed surface, or set of closed surfaces if multimodality is present. Scott (1992, Section 1.4) describes this approach and explores its use on a variety of datasets.

Figure 1.9 displays a contour plot of a three-dimensional density estimate of the geyser data, using normal kernel functions. It is represented as a 'wire frame' object, constructed in a manner similar to that described by Scott (1992, Appendix A). This contour has been chosen, as described in the previous section, to enclose exactly 75% of the observations. The space contained by the contour therefore corresponds to the upper reaches of the density estimate. This focuses

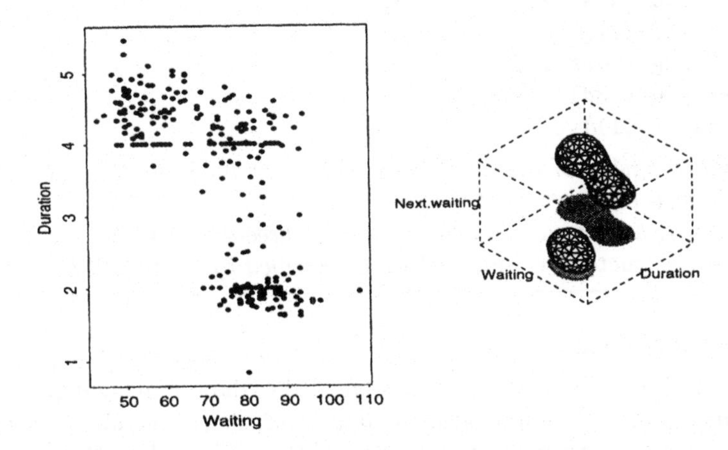

FIG. 1.9. A contour plot of the three-dimensional density estimate of the geyser data.

attention on the principal features of the density and ignores the outer regions, where behaviour is influenced by small numbers of observations.

The density contour of the geyser data is disjoint, confirming the highly clustered nature of the data. If a smaller percentage had been specified for the fraction of data contained, three disjoint surfaces would have been apparent, corresponding to the three clusters in the data. However, the contour displayed in the figure draws attention to the fact that the two upper clusters are not as well separated as the lower one is from the other two.

When surfaces are disjoint the relative locations of the separate parts in three dimensions may not always be clear. This has been resolved in Fig. 1.9 by displaying the shadow of the surfaces on the floor of the plot, as if a light were shining from the top face of the surrounding cube.

Scott (1992) describes a variety of more sophisticated techniques for constructing and displaying density contours, including the use of light sources to enhance visual perception, and the removal of strips of the surface in order to display interior contours.

Density estimation can be carried out in four, five and more dimensions. However, this should be done with care as there are increasing problems due to the 'curse of dimensionality', as described by Scott (1992, Chapter 7).

S-Plus Illustration 1.7. A three-dimensional density contour plot

---

Figure 1.9 was constructed with the following S-Plus *code. The geyser data are supplied in* S-Plus *as a standard dataset. However, the data used here have been organised in a manner which is more convenient for a three-dimensional analysis.*

The calculations for the density estimate may take some considerable time on some computers.

```
provide.data(geys3d)
par(mfrow=c(1,2))
plot(Waiting, Duration)
sm.density(geys3d)
par(mfrow=c(1,1))
```

The original version of the function to plot three-dimensional contours was written by Stuart Young.

More sophisticated forms of visualisation of three-dimensional density contours can be constructed by S-Plus *software described by Scott (1992).*

1.5 Directional data

The concept of density estimation can be extended to data of many other types. A more unusual example is provided by data which define directions in three dimensions, where the sample space consists of the surface of a sphere. Fisher *et al.* (1987) give an excellent introduction to this type of data and discuss in particular a variety of means of graphical exploration. One example concerns data on magnetic remanence. For each specimen, angles of latitude and longitude are measured and the data can therefore be represented as points on the surface of the unit sphere. Several different transformations are available to map the surface of the sphere onto a plane. These transformations have a variety of properties such as preserving areas or angles. For visual inspection there are some advantages in representing the sphere as a three-dimensional object, which is referred to as the 'orthographic' projection. The eye is used to viewing spheres, as in the globes which provide models of the earth, and lines of latitude and longitude can be marked to indicate the curved shape of the sphere. Figure 1.10 shows a display of this type for the magnetic remanence data. The two panels represent the front and rear views of the sphere.

Fisher *et al.* (1987) also propose the use of a density estimate in order to highlight the main features of the distribution. Diggle and Fisher (1985) employed density estimate contours, using a Fisher density for the kernel function, of the form

$$\hat{f}(y_1, y_2, y_3) \propto \sum_{i=1}^{n} \exp\{\kappa(y_1 y_{1i} + y_2 y_{2i} + y_3 y_{3i})\},$$

where $\{(y_{1i}, y_{2i}, y_{3i}); i = 1, \ldots, n\}$ denote the observations in three-dimensional co-ordinates as points on the sphere and κ plays the role of a smoothing parameter. Since it is the relative heights of the density estimate which are important, the constants necessary to ensure integration to 1 have been ignored. The left panel of Fig. 1.11 shows the shape of the kernel function in this case. A smoothing parameter $\kappa = 13.6$ has been used, as recommended by Fisher *et al.* (1987). The shape of the density function is displayed by associating density height with the intensity of the grey scaling. This reveals the underlying pattern in the data quite effectively.

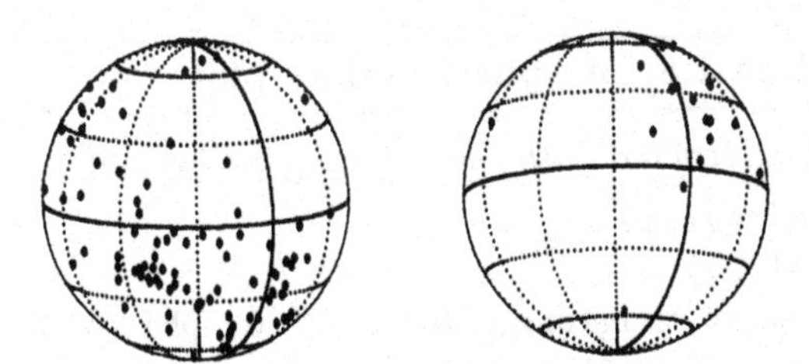

FIG. 1.10. A spherical data plot of the magnetic remanence data. The left panel shows the front view and the right panel shows the rear view.

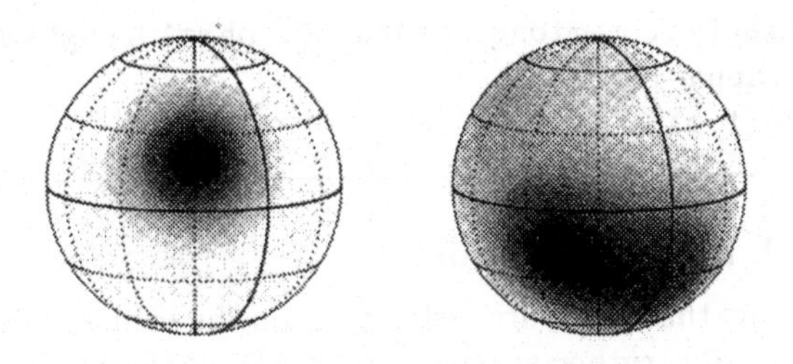

FIG. 1.11. A kernel function (left) and a spherical density estimate (right) from the magnetic remanence data.

A simpler form of directional data consists of single angular measurements which arise, for example, with quantities such as wind direction. Here the data, when recorded as radians, are restricted to the range $[0, 2\pi]$, with the additional feature that the end points of this range coincide. A natural model in this setting is the von Mises distribution, whose density function in θ is proportional to $\exp\{\kappa \cos(\theta - \theta_0)\}$, where $\theta, \theta_0 \in [0, 2\pi]$, and so this also forms a natural choice of kernel function when constructing a density estimate for circular data.

A similar effect can be achieved by replicating the data in $[0, 2\pi]$ over the adjoining ranges $[-2\pi, 0]$ and $[2\pi, 4\pi]$ to create the circularity in the contribution of each observation, while retaining the factor $1/n$ for the original sample size, as described by Silverman (1986, Section 2.10).

S-Plus Illustration 1.8. A spherical data plot

Figure 1.10 was constructed with the following S-Plus *code.*

```
provide.data(magrem)
par(mfrow=c(1,2))
sm.sphere(maglat, maglong, theta =  60, phi =  10)
sm.sphere(maglat, maglong, theta = 240, phi = -10)
par(mfrow=c(1,1))
```

An original version of the function `sm.sphere` *for plotting spherical data was written with the help of Adrian Hines.*

S-Plus Illustration 1.9. A spherical density estimate

Figure 1.11 was constructed with the following S-Plus *code.*

```
provide.data(magrem)
par(mfrow=c(1,2))
sm.sphere(20, -30, theta=60, phi=10, sphim=T, kappa=13.6)
sm.sphere(maglat, maglong, theta=60, phi=10, sphim=T,
          kappa=13.6)
par(mfrow=c(1,1))
```

1.6 Data with bounded support

It commonly occurs that there are restrictions on the values which data can take. Directional data are a rather special case of this. A more common example is where only positive values can be recorded. All the variables in the aircraft data are of this type. In view of the strong skewness which exists in these data the underlying distributional patterns are best viewed on a log scale. However, it is also of some interest to present the density functions on the original scale.

Figure 1.12 shows a histogram and density estimate of the aircraft speed data, for all years. One problem with the density estimate is that the kernel functions centred on the observations which are very close to zero have transferred positive weight to the negative axis. This effect could be reduced by employing a smaller smoothing parameter, but this would have unwelcome effects elsewhere in the estimate, and in particular the size of the variations which already appear in the long right hand tail of the estimate would be increased.

At least two approaches to the problem are possible. In the ***transformation method*** the variable Y can be transformed to a new variable $t(Y)$ with unbounded support, the density of $t(Y)$ estimated and then transformed back to the original scale. In practice, the whole process can be accomplished in one operation. If

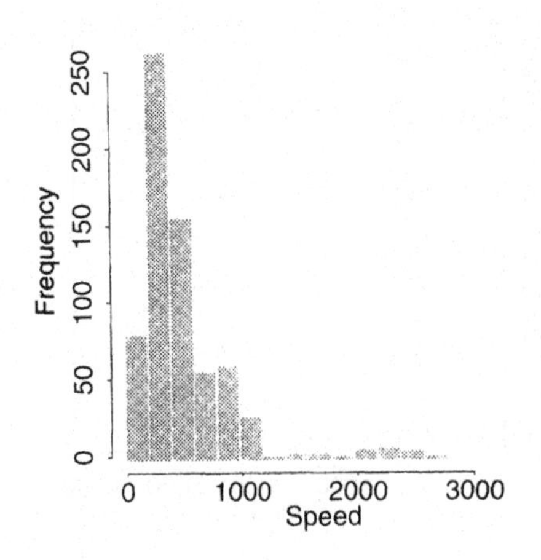

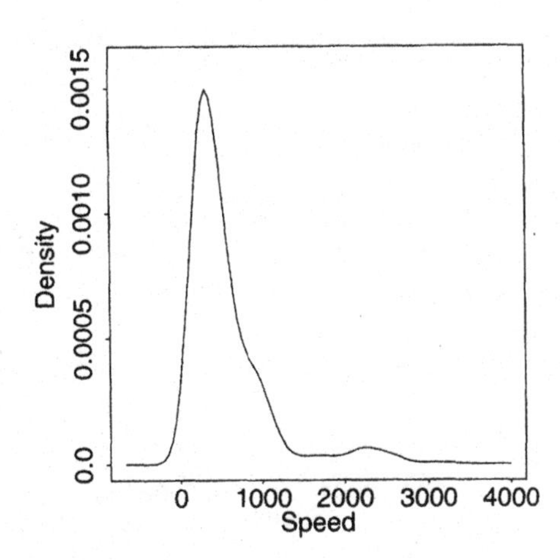

FIG. 1.12. A histogram and density estimate of the speed data on its original scale.

g denotes the probability density function of $t(Y)$, then the two densities are related by

$$f(y) = g\,(t(y))\; t'(y)$$

and $g(\cdot)$ can be estimated by the standard method, leading then to the estimate

$$\hat{f}(y) = \frac{1}{n} \sum_i w(t(y) - t(y_i); h)\, t'(y).$$

Wand and Jones (1995, p.43) describe this approach in some detail.

In the *modified kernel* method the kernel function w can be modified in a suitable manner. For estimation of f at a point y, a kernel function can be chosen so that its support does not exceed the support of Y. A related approach is to use standard kernel functions but to modify these near a boundary. A family of 'boundary kernels' was proposed by Gasser and Müller (1979) and is also described in Wand and Jones (1995, Section 2.11).

The effect of the transformation approach, using the log function, is illustrated in the left panel of Fig. 1.13 with the aircraft speed data. The density estimate now lies only on the positive section of the axis, and in addition the variations in the right hand tail have been eliminated. In this case the transformation method provides a satisfactory solution, and in general it does not often seem to be necessary to use other approaches, such as modified kernels. Clearly the transformation method also applies to other forms of restricted data. For example, when the support of Y is a bounded interval (a, b) then a natural transformation to consider is

$$t(x) = \log \frac{x-a}{b-x},$$

which maps (a, b) to the real line.

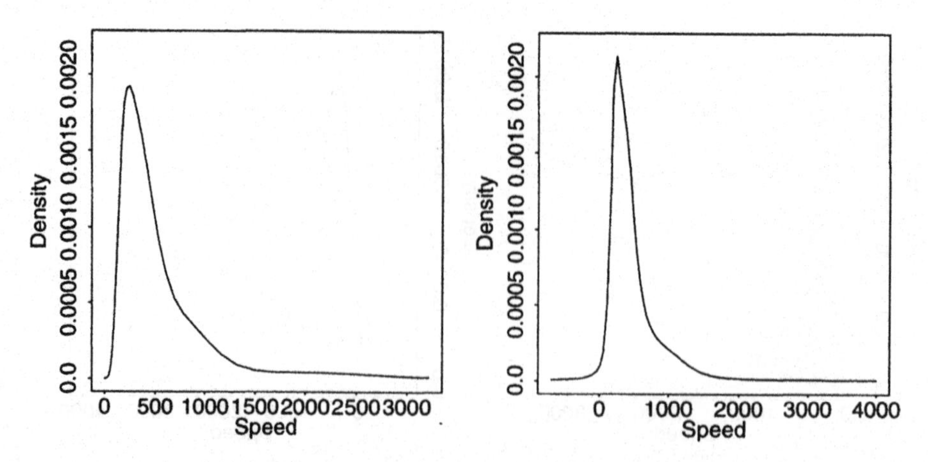

FIG. 1.13. Density estimates based on the aircraft speed data. The left panel is based on the log transformation method. The right panel is based on the variable kernel method, discussed in Section 1.7, with the nearest neighbour parameter $k = 30$.

Wand *et al.* (1991) discuss the interrelationships of the shape of the original distribution, the use of different transformation families and the selection of appropriate transformation parameters, on the performance of the final estimate.

Positive data in two dimensions can be handled in a similar way. This is explored in an exercise at the end of the chapter.

S-Plus Illustration 1.10. A density estimate from the speed data

Figure 1.12 was constructed with the following S-Plus *code. The* `yht` *parameter controls the height of the vertical axis in the plot of the density estimate.*

```
provide.data(aircraft)
par(mfrow=c(1,2))
hist(Speed, ylab="Frequency")
sm.density(Speed, yht=0.0016)
par(mfrow=c(1,1))
```

S-Plus Illustration 1.11. Density estimates for bounded support

Figure 1.13 was constructed with the following S-Plus *code. The function* `nnbr` *is provided in the* `sm` *library. It returns a vector consisting of the nearest neighbours from each observation to the remainder of the data. This feature of density estimation is discussed in Section 1.7.*

```
provide.data(aircraft)
hw <- nnbr(Speed, 30)
hw <- hw/exp(mean(log(hw)))
```

```
par(mfrow=c(1,2))
sm.density(Speed, yht=0.0022, positive=T)
sm.density(Speed, yht=0.0022, xlim=c(-700,4000), h.weights=hw)
par(mfrow=c(1,1))
```

1.7 Alternative forms of density estimation

Throughout this book the kernel approach is used because it is conceptually simple, deriving naturally from the histogram in the case of density estimation, and computationally straightforward. The techniques also extend naturally to the multivariate case and there are close links with techniques based on kernel functions for the smoothing of regression data. However, there are other approaches to density estimation and some of these are sketched below. The first is a variant of the kernel approach which is particularly useful for certain kinds of data. The others take rather different routes in constructing estimators and, after a brief description, will not be pursued further.

1.7.1 *Variable bandwidths*

One of the features of the speed data, illustrated in Fig. 1.12, is the presence of 'bumps' in the right hand tail of the distribution, caused by small numbers of observations. Here the data are sparse and it may be more appropriate to use a large smoothing parameter to remove these bumps in the estimate. On the other hand, where the data are clustered closely together at small values of speed it would be appropriate to use a small smoothing parameter. This idea has led to a variety of suggestions for *variable* bandwidths, where a different smoothing parameter can be used in each kernel function, of the form

$$\hat{f}(y) = \frac{1}{n}\sum_{i=1}^{n} w(y - y_i; h_i).$$

One possibility is to reflect the sparsity of the data in expressions such as $h_i = hd_k(y_i)$, where h denotes an overall smoothing parameter and $d_k(y_i)$ denotes the distance from y_i to its kth nearest neighbour among the data. Breiman *et al.* (1977) describe this approach. Silverman (1986) makes the helpful suggestion of introducing the variable element as a modification of the overall smoothing parameter, to give $h_i = hd_k(y_i)/\bar{d}$, where $\bar{d}$ denotes the geometric mean of the $d_k(y_i)$.

The right panel of Fig. 1.13 shows an estimator of this type, using $k = 30$ nearest neighbours. The behaviour in the right hand tail has been improved, as the kernel functions there are much flatter. However, this estimator has not entirely overcome the problem of the transfer of positive weight to the negative axis discussed in the previous section.

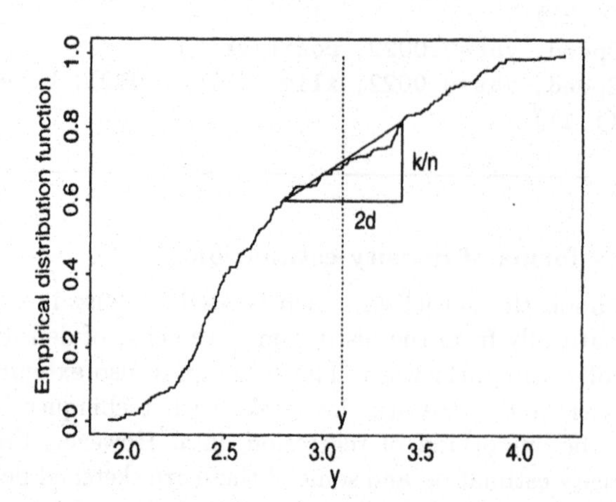

FIG. 1.14. A density estimate as the gradient of the empirical distribution function, using the aircraft span data, on a log scale, for the third time period.

1.7.2 *Nearest neighbour methods*

Figure 1.14 displays the empirical distribution function of the aircraft span data, on a log scale, for the third time period. Since a density can be obtained by finding the derivative of the distribution function, a density estimate can be constructed by measuring the gradient of the curve, as indicated graphically in this figure.

This can be done in two ways. The first is to fix a distance d on either side of the point of estimation y. This is equivalent to the use of the density estimator (1.1) with a 'box' kernel defined as $I(z;d)/(2d)$, where $I(z;d)$ is the indicator function of the interval $(-d,d)$. The density estimate is then given by $(k/n)/(2d)$, where k represents the number of observations lying within a distance d of y, and whose kernel function therefore contributes to the density estimate at that point.

An alternative approach is to fix the value of k, the number of observations which contribute to the density estimate. This then determines d, which is referred to as the kth nearest neighbour distance. The nearest neighbour form of density estimate is therefore available as $\tilde{f}(y) = (k/n)/(2d_k(y))$, using the full notation for the nearest neighbour distance $d_k(y)$.

Moore and Yackel (1977) generalised work by Loftsgaarden and Quesenberry (1965) to construct a more general *nearest neighbour* density estimator of the form

$$\hat{f}(y) = \frac{1}{n}\sum w(y - y_i; hd_k(y)).$$

The distinctive feature of this is that the bandwidth changes with the point of estimation y rather than with the observation y_i. An unfortunate consequence is that the estimator does not necessarily integrate to 1. In addition, there can be abrupt changes in nearest neighbour distances as a function of y and so the resulting density estimate can display a lack of smoothness.

Both nearest neighbour and variable bandwidth methods extend to the multivariate case since a nearest neighbour distance is easily computed in several dimensions.

1.7.3 *Orthogonal series methods*

To introduce this approach, it is useful to recall the concept of a Fourier series. A function f defined on $(-\pi, \pi)$ can be represented by the series expansion

$$f(x) = \tfrac{1}{2}a_0 + \sum_{k=1}^{\infty}(a_k \cos kx + b_k \sin kx), \tag{1.2}$$

where the Fourier coefficients a_k, b_k are defined by

$$a_k = \frac{1}{\pi}\int_{-\pi}^{\pi} f(x)\cos(kx)dx, \quad b_k = \frac{1}{\pi}\int_{-\pi}^{\pi} f(x)\sin(kx)dx.$$

This construction and the discussion below require the technical condition of square integrability of the functions involved, but we shall not examine these mathematical aspects in further detail.

Equation (1.2) expresses $f(x)$ as a linear combination of the set of trigonometric functions

$$\{1,\ \cos x,\ \sin x,\ \cos 2x,\ \sin 2x,\ \ldots\},$$

which are then said to form a ***basis*** for the set of possible functions $\{f\}$. By changing the coefficients of the linear combination, different functions f are generated.

This scheme can be implemented with the trigonometric basis replaced by some other set of functions $\{\psi_k(x); k = 0, 1, \ldots\}$. An extension of this kind is required to handle functions defined over an unrestricted set X, such as the real line, since the Fourier series allows only periodic functions. The generalisation of the Fourier series is then

$$f(x) = \sum_{k=0}^{\infty} c_k \psi_k(x) \tag{1.3}$$

where the coefficients are defined by

$$c_k = \int_X \psi_k(x)\, f(x)\, w(x) dx$$

and $w(x)$ is a weight function on X. However, not all sets of functions $\{\psi_k(x)\}$ are suitable for this purpose. They need to form an orthogonal basis, which means that:

1. $\int_X \psi_k(x)\psi_j(x)w(x)dx = 0$ if $k \neq j$;

2. the set of functions generated by linear combinations of the ψ_k must include 'all' functions f defined on X.

The discussion so far has been in terms of mathematical analysis. The usefulness of these concepts for the problem of nonparametric density estimation was pointed out by Čencov (1962). In this context, f denotes the density function of a random variable Y, and the coefficients c_k have a direct interpretation as expected values, namely

$$c_k = \mathbb{E}\{\psi_k(Y)w(Y)\}.$$

This allows a simple method of estimation of the coefficients. If a random sample $\{y_1, \ldots, y_n\}$ is drawn from Y, then an estimate of c_k is given by

$$\hat{c}_k = \frac{1}{n}\sum_{i=1}^{n} \psi_k(y_i)w(y_i).$$

By substituting this value into (1.3), an estimate of f is produced. This has the form

$$\hat{f}_P(x) = \sum_{k=1}^{m} \hat{c}_k \psi_k(x)w(x) \tag{1.4}$$

where the number m of summands is finite, in contrast with (1.3).

There is a broad analogy between the kernel density estimate and the *projection* (or orthogonal series) estimate (1.4), in the sense that, in both cases, there are two ingredients to be chosen. For the kernel methods these are the kernel function and the bandwidth h. For the projection method they are the basis $\{\psi_k(x)\}$ and the number of terms m. The analogy between h and m is particularly close. Increasing m in (1.4) has the same effect as decreasing h in (1.1), namely it makes the estimate more wiggly, while decreasing m makes the curve smoother. Moreover, it can be shown that m must increase with n but at a slower rate, for the best performance. This again parallels the behaviour of h, but in the reverse direction, as will become apparent in Chapter 2.

Additional mathematical aspects of this estimator have been studied by Schwartz (1967). An estimator of the form (1.4) has been studied in detail by Kronmal and Tarter (1968), for the case when the support of the variable Y is a finite interval and the chosen basis is the trigonometric one.

One advantage of (1.4) over the kernel methods, and also others, is that it produces a 'proper' function. Once the $\hat{c}_k$ have been computed, they represent the estimated function over the whole range of possible values. A disadvantage of the method is that some people find its construction less intuitive. Moreover, (1.4) can occasionally take negative values in the tails of the distribution.

1.7.4 *Local likelihood and semiparametric density estimation*

A recent interesting approach to density estimation is based on the idea of local likelihood. This allows a model to be fitted locally by assigning weights to each observation according to their distance from some point of interest. Copas (1995) and Hjört and Jones (1996) describe this method, which can be considered as a semiparametric form of density estimation. The approach has a similar philosophy to some methods of nonparametric regression which will be described in Chapter 3.

1.8 Further reading

There are several excellent texts on density estimation, each with its own particular emphasis. Silverman (1986) provides a very accessible introduction. Wand and Jones (1995) add much new research and provide a helpful link with nonparametric regression. Scott (1992) develops the multivariate approach, emphasising visualisation in particular and, among many other aspects, describing more sophisticated, and highly efficient, histogram based estimators. Simonoff (1996) illustrates many aspects of density estimation, and in particular provides a very extensive bibliography. The comments below provide some pointers to the applications of smoothing techniques in other, related areas.

The empirical distribution function is a standard tool for displaying samples of data. Its representation as a step function has the advantage that individual observations can be identified from the plot. However, it is often reasonable to assume that the true underlying distribution function is smooth. Motivations for constructing a smooth estimate include the more attractive visual appeal of a smooth curve and the potential for improved estimation.

By analogy with density estimation, a smooth estimate can be constructed by placing a kernel function, which is itself a distribution function, over each observation in the form

$$\hat{F}(y) = \frac{1}{n} \sum_{i=1}^{n} W(y - y_i; h),$$

where the parameter h controls the standard deviation associated with the distribution function W, and hence controls the degree of smoothing applied to the data. The standard empirical distribution function is obtained by letting $h \to 0$, when W is replaced by the indicator function

$$I(y - y_i) = \begin{cases} 1 & \text{if } y - y_i \geq 0 \,, \\ 0 & \text{otherwise.} \end{cases}$$

Notice that the estimator $\hat{F}$ can also be viewed as the integral of the density estimate (1.1). It turns out that a smaller degree of smoothing is appropriate for the distribution function, compared to the density, because the integral is a smoother function that the integrand.

Smooth estimation for distribution functions was first proposed by Nadaraya (1964a). More recent work includes Azzalini (1981), Falk (1983) and Fernholz (1991).

When some observations in a sample are censored, a convenient approach to estimating the distribution function is through the Kaplan–Meier survivor function. This leads to an estimate of the underlying distribution function with the distinctive feature that the steps are not all of size $1/n$. This suggests that in the case of censored data a density estimate should be constructed of the form

$$\hat{f}_c(y) = \sum_{i=1}^{n} \nu_i w(y - y_i; h),$$

where ν_i denotes the size of the jump in the Kaplan–Meier estimator at observation y_i. Note that if y_i is censored then the size of the jump is 0. Wand and Jones (1995, Section 6.2.3) give some details of this approach.

For lifetime data, the hazard function $f(y)/\{1 - F(y)\}$ offers a useful alternative characterisation of a distribution. Watson and Leadbetter (1964) proposed that an estimate be constructed from separate estimates of f and F. Silverman (1986, Section 6.5.1) illustrates this approach, and recent work is reported by Rosenberg (1995).

In a two-dimensional setting, the analysis of point process data leads to interest in the underlying intensity function. In fact, conditional on the number of observations in the sample, this is equivalent to the density function and so there is a very close connection between the two. Diggle (1985) discusses kernel methods for this type of data. An example of spatial point patterns will be discussed in Chapter 6.

Compositional data, where each observation consists of a list of proportions, form another example where the sample space consists of a bounded region. Aitchison and Lauder (1985) describe how density estimation can be conducted in this setting, using kernels based on the logistic normal density function.

Data from multivariate discrete distributions often lead to small numbers of observations in each cell. Kernel methods of smoothing discrete data were developed by Aitchison and Aitken (1976) and more recent work is described by Simonoff (1996).

Efficient computation of density estimates is required when sample sizes are large. This can be achieved by binning techniques, which are described in some detail by Wand and Jones (1995, Appendix D).

Exercises

1.1 *Aircraft data.* The aircraft span data for the third time period were explored in Section 1.2. Repeat this for aircraft `Speed`, also on a log scale. Is a density estimate more informative than a simple histogram? Explore the effect of changing the smoothing parameter and consider whether the distribution is multimodal. Repeat this for the variables `Power`, `Length`, `Weight` and `Range`. Compare some of these variables over time by superimposing density estimates for the three different periods.

1.2 *Aluminium oxide data.* The file `Al2O3` contains data on the percentages of aluminium oxide found in geological samples. The data are discussed in Chapter 2, where some details on the context of the study are provided.

Construct density estimates from the data and consider whether there is any evidence that the data are not normally distributed.

1.3 *Simulated data.* Use the `rnorm` function of S-Plus to generate repeated sets of normally distributed data and view the variation which occurs in the density estimates. Repeat this for other distributional shapes.

1.4 *Kernel function shape.* The S-Plus function `ksmooth` allows density estimates to be constructed from different shapes of kernel functions. For example, the commands

```
y <- rnorm(50)
plot(c(-4,4), c(0,0.4), type="n")
lines(ksmooth(y, bandwidth=1))
```

will plot a density estimate using a 'box' kernel by default. Compare the effects of using other kernel shapes, with the same value of bandwidth, by adding to `ksmooth` the argument `kernel` with the values `"triangle"`, `"parzen"` and `"normal"`.

1.5 *Circular data.* Write a short piece of S-Plus code which will create a density estimate over the range $(0, 2\pi)$ from a vector of data which corresponds to angular measurements. Use the technique of replicating the data over the intervals $(-2\pi, 0)$ and $(2\pi, 4\pi)$, mentioned in Section 1.5, to ensure that the density estimate takes the same value at 0 and 2π. (Readers who enjoy a challenge might like to produce a plot of the density estimate on a unit circle.)

1.6 *Bivariate views of the aircraft data.* In Section 1.3 a density estimate of the first two principal components of the aircraft data, for the third time period, was found to be trimodal. Produce density estimates for the first two time periods and explore whether multimodality is also present there.

1.7 *The effect of bandwidth on contours.* The density contours described in Section 1.3 are designed to contain specific proportions of the data. It is therefore the *relative*, rather than absolute, density heights at the observations which are important in defining the contours. Explore the effect of changing the bandwidth, through the `hmult` parameter, on the contours of the first two principal components of the aircraft data.
Prove that as the smoothing parameter increases in the two-dimensional case, the ranking of the observations by estimated density height converges to the same ranking as that imposed by a fitted bivariate normal distribution.
Look at the effect of using very large smoothing parameters to construct density contours from the aircraft data.

1.8 *Two-dimensional discrimination.* Ripley (1994) discussed the performance of neural networks and related methods of classification. A test example used data which were simulated from equal mixtures of bivariate normal distributions. One group had a component centred at $(-0.7, 0.3)$ and another at $(0.3, 0.3)$. A second group had components centred at $(-0.3, 0.7)$

and $(0.4, 0.7)$. All the components had variance 0.03 in each direction, and covariance 0. Use a sample size of your own choice, with equal numbers in each group, to simulate data from each of these distributions. Estimate a density function for each group, using identical evaluation grids, by setting the parameters `xlim` and `ylim` in `sm.density`. The output of the `sm.density` function has a component `estimate` which can be used to extract the estimate numerically. Calculate the difference of the two estimates and superimpose the 0 contour of this difference on a plot of the data, to provide a graphical discrimination line between the two groups.

1.9 *Pole position.* Fisher *et al.* (1987) provide data on the historical positions of the South Pole, from a palaeomagnetic study of New Caledonia laterites. The measurements are recorded as latitude and longitude positions and are available in the `poles` file. Use `sm.sphere` to plot the data, using a suitable orientation. Construct a density estimate and identify its mode graphically, as an estimate of the location of the South Pole.

1.10 *Variable bandwidths.* Examine the effect of changing the nearest neighbour parameter k in a variable bandwidth estimate of the aircraft speed data, while holding the overall parameter h fixed.

1.11 *Nearest neighbour density estimate.* Write a function which will construct the nearest neighbour form of density estimate described in Section 1.7.

1.12 *Data restricted to* $[0, 1]$. Generate a sample of data from a beta distribution. Apply the logistic transformation $\log(p/(1-p))$, use `sm.density` to estimate the density function on that scale and transform the result back to the original scale. Compare your estimate with a histogram of the data.

1.13 *Censored data.* A density estimate for the case where the values of some observations in a sample are censored was described in Section 1.8. This requires weights determined by the step sizes of the Kaplan–Meier survivor function. These can be derived from the S-Plus function `survfit`. Construct a density estimate from censored data of your own choice by passing these weights to `sm.density` through the `weights` argument. Consider what happens when the largest observation in a sample is censored.

1.14 *Positive data in two dimensions.* Use a suitable distribution, such as a product of two gamma densities, to generate two-dimensional data with positive components. Use `sm.density` with the argument `positive=T` to explore the effectiveness of the transformation method, described in Section 1.6, in two dimensions.

2

DENSITY ESTIMATION FOR INFERENCE

2.1 Introduction

In Chapter 1, the role of kernel functions in constructing nonparametric density estimates was illustrated on a variety of different types of data. This led to a set of useful tools, particularly for graphical illustration of distributional shape. The uses of density estimates in reaching conclusions in a more quantitative way will now be investigated. Attention will be focused on one- and two-dimensional data on an unrestricted continuous scale.

In order to go beyond the exploratory and graphical stage it is necessary first to understand more about the behaviour of these estimators and to derive some basic properties. Although many theoretical results exist, simple expressions for means and variances of the estimators are enough to allow ideas of interval estimation and hypothesis testing to be discussed, and to motivate techniques for choosing an appropriate bandwidth to employ with a particular dataset.

2.2 Basic properties of density estimates

A simple manipulation shows that the mean of the density estimator (1.1) can be written as

$$\mathbb{E}\left\{\hat{f}(y)\right\} = \int w(y-z;h)f(z)dz. \tag{2.1}$$

This is a convolution of the true density function f with the kernel function w. Smoothing has therefore produced a biased estimator, whose mean is a smoothed version of the true density. Further insight can be gained through a Taylor series argument. It is convenient to use a kernel function which, with a slight change of notation, can be parametrised in the form $(1/h)w(z/h)$. A Taylor series expansion then produces the approximation

$$\mathbb{E}\left\{\hat{f}(y)\right\} \approx f(y) + \frac{h^2}{2}\sigma_w^2 f''(y), \tag{2.2}$$

where σ_w^2 denotes the variance of the kernel function, namely $\int z^2 w(z)dz$. Since $f''(y)$ measures the rate of curvature of the density function, this expresses the fact that $\hat{f}$ underestimates f at peaks in the true density and overestimates at troughs. The size of the bias is affected by the smoothing parameter h. The component σ_w^2 will reduce to 1 if the kernel function w is chosen to have unit variance.

Through another Taylor series argument, the variance of the density estimate can be approximated by

$$\mathrm{var}\left\{\hat{f}(y)\right\} \approx \frac{1}{nh} f(y)\,\alpha(w), \tag{2.3}$$

where $\alpha(w) = \int w^2(z)dz$. As ever, the variance is inversely proportional to sample size. In fact, the term nh can be viewed as governing the local sample size, since h controls the number of observations whose kernel weight contributes to the estimate at y. It is also useful to note that the variance is approximately proportional to the height of the true density function.

These approximate expressions for the mean and variance of a density estimate encapsulate the effects of the smoothing parameter which were observed in Fig. 1.3. As h decreases, bias diminishes while variance increases. As h increases the opposite occurs. The combined effect of these properties is that, in order to produce an estimator which converges to the true density function f, it is necessary that both h and $1/nh$ decrease as the sample size increases. A suitable version of the central limit theorem can also be used to show that the distribution of the estimator is asymptotically normal.

A similar analysis enables approximate expressions to be derived for the mean and variance of a density estimate in the multivariate case. In p dimensions, with a kernel function defined as the product of univariate components w, and with smoothing parameters $(h_1, \ldots, h_p)$, these expressions are

$$\mathbb{E}\left\{\hat{f}(y)\right\} \approx f(y) + \frac{1}{2}\sigma_w^2 \left[\sum_{j=1}^{p} h_j^2 \frac{\partial^2}{\partial y_j^2} f(y)\right],$$

$$\mathrm{var}\left\{\hat{f}(y)\right\} \approx \frac{1}{nh_1 \cdots h_p} f(y)\,\alpha(w)^p.$$

Wand and Jones (1995, Section 4.3) derive results for more general kernel functions.

It is helpful to define an overall measure of how effective $\hat{f}$ is in estimating f. A simple choice for this is the *mean integrated squared error* (MISE) which, in the one-dimensional case, is

$$\begin{aligned}\mathrm{MISE}(\hat{f}) &= \mathbb{E}\left\{\int [\hat{f}(y) - f(y)]^2 dy\right\} \\ &= \int \left[\mathbb{E}\left\{\hat{f}(y)\right\} - f(y)\right]^2 dy + \int \mathrm{var}\left\{\hat{f}(y)\right\} dy.\end{aligned}$$

This combination of bias and variance, integrated over the sample space, has been the convenient focus of most of the theoretical work carried out on these estimates. In particular, the Taylor series approximations described in (2.2) and (2.3) allow the mean integrated squared error to be approximated as

$$\text{MISE}(\hat{f}) \approx \frac{1}{4} h^4 \sigma_w^4 \int f''(y)^2 dy + \frac{1}{nh} \alpha(w).$$

Establishing the properties of the estimators which employ variable bandwidths, as described in Chapter 1, is more complex. Here the smoothing parameter h in the kernel function over observation y_i is replaced by expressions such as $hd_k(y_i)$, where $d_k(y_i)$ denotes the distance from y_i to its kth nearest neighbour among the data. It was shown in Section 1.7 that a nearest neighbour distance is inversely proportional to a simple form of density estimate. In view of this, it is instructive to represent smoothing parameters involving variable bandwidths as $h/\tilde{f}(y_i)$, where $\tilde{f}$ denotes a density estimate. This shows that this approach is based on a *pilot* estimate of the underlying density which is then used to adjust the kernel widths locally. This representation also suggests a means of investigating the behaviour of estimators of this type by analysing estimators which use the true density f as a pilot estimator. Abramson (1982) adopted this approach and used Taylor series expansions to derive first-order asymptotic properties. An additional exploration of more general variable bandwidths of the form $h/\tilde{f}(y_i)^\alpha$ led to the proposal that the power $\alpha = 1/2$ is most appropriate, since the asymptotic arguments then suggest that the principal bias term is eliminated. This results also holds in the multivariate case.

Bowman and Foster (1993) avoided asymptotic calculations by employing numerical integration in the calculations of mean and variance. This showed that although considerable caution should be exercised with Taylor series approximations in this context, the broad conclusions of these analyses were corroborated.

As an illustration of variable bandwidths, Fig. 2.1 displays density estimates constructed from data on a tephra layer, resulting from a volcanic eruption in Iceland around 3500 years ago. Dugmore *et al.* (1992) report the geological background to the collection of these data, which refer to the chemical composition of shards of volcanic glass found in the tephra layer. These compositions can help to identify whether tephra layers have resulted from the same eruption. Figure 2.1 displays data on the percentage of aluminium oxide (Al_2O_3) in the sample from a single site. A natural model for compositional data is to apply a logistic transformation as described by Aitchison (1986).

Three density estimates have been produced in Fig. 2.1. In each case the same smoothing parameter was used, but in one of the estimates nearest neighbour weights were used, and in another these weights were calculated on a square root scale. In both the estimates involving variable kernels the weights were scaled, as described in Section 1.7, in order to allow comparisons to be made. The main effect of the variable weights is to allow the density estimate to peak more sharply. This peak is more pronounced for the nearest neighbour weights, corresponding to a pilot density estimate $\hat{f}$. The square root scale offers a more modest modification which has some backing from the theory of Abramson (1982).

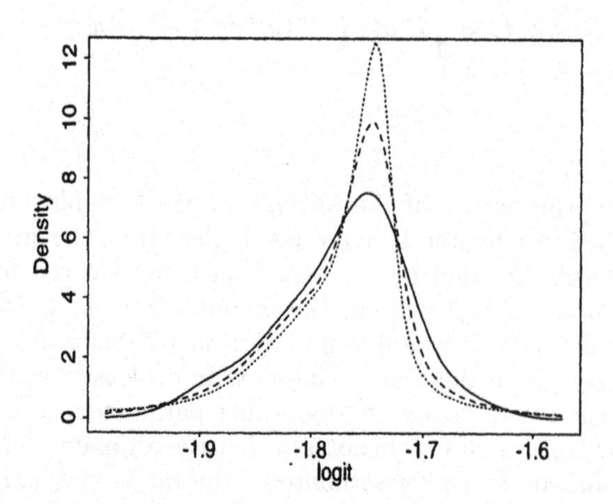

FIG. 2.1. Density estimates from the tephra data, on a logit scale, using simple kernels (full line), nearest neighbour weights (dotted line) and square root nearest neighbour weights (dashed line).

S-Plus Illustration 2.1. Square root weights for variable kernels

Figure 2.1 was constructed with the following S-Plus *code.*

```
provide.data(tephra)
logit <- log(Al2O3/(100-Al2O3))
nn    <- nnbr(logit, 7)
hw    <- nn/exp(mean(log(nn)))
sm.density(logit, h.weights = hw, lty = 2, yht = 12.2)
hw    <- sqrt(nn)
hw    <- hw/exp(mean(log(hw)))
sm.density(logit, h.weights = hw, lty = 3, add = T)
sm.density(logit, add = T)
```

Mathematical aspects: The approximate mean and variance of a one-dimensional density estimate

An asymptotic argument for the mean of a density estimate begins with (2.1). A change of variable produces the expression

$$\mathbb{E}\left\{\hat{f}(y)\right\} = \int w(z)\, f(y - hz)\, dz.$$

If $f(y - hz)$ is now replaced by the Taylor series expansion $f(y) - hzf'(y) + \frac{1}{2}h^2z^2f''(y)$, then the principal terms allow the mean to be approximated as

$$\mathbb{E}\left\{\hat{f}(y)\right\} \approx f(y) + \frac{h^2}{2}\int z^2 w(z) dz\, f''(y),$$

where the remainder term in this approximation is $o(h^2)$. The term involving h reduces to 0 because $\int zw(z)dz = 0$.

The variance can be written as

$$\begin{aligned}\text{var}\left\{\hat{f}(y)\right\} &= \frac{1}{n}\text{var}_z\{w(y-z\,;h)\} \\ &= \frac{1}{n}\{\mathbb{E}_z\left\{w(y-z\,;h)^2\right\} - \mathbb{E}_z\left\{w(y-z\,;h)\right\}^2\}.\end{aligned}$$

The expectations in this last expression can be expanded by a change of variable and Taylor series, as above, to yield the approximate expression (2.3).

2.3 Confidence and variability bands

One of the issues raised in Chapter 1 by the use of density estimates in exploring data was the need to assess which features of a density estimate indicate genuine underlying structure and which can be attributed to random variation. It is best to construct tools for this purpose around specific questions. For example, methods for assessing normality, and the independence of variables in a bivariate distribution, will be derived later in this chapter. However, it would also be of some value to have a general purpose method of indicating the uncertainty associated with a density estimate.

A confidence band, displaying confidence intervals for the true density at a range of values of y, would be very helpful. The immediate difficulty is the forms of the mean and variance of a density estimate, outlined in Section 2.2. The variance involves the true density $f(y)$, and the mean shows that bias is present, involving the second derivative $f''(y)$. In theory it is possible to use estimates of these unknown factors. However, this introduces an unsatisfactory degree of complexity, and further uncertainty, into the problem.

The bias component is particularly problematic, through the involvement of $f''(y)$. This leads to the idea of addressing the lesser, but still useful, aim of quantifying the variability of a density estimate without attempting to account for bias. Since the expression for variance has a relatively simple form, a variance stabilising argument can be adopted. For any transformation $t(\cdot)$, a Taylor series argument shows that

$$\text{var}\left\{t(\hat{f}(y))\right\} \approx \text{var}\left\{\hat{f}(y)\right\}\left[t'\left(\mathbb{E}\left\{\hat{f}(y)\right\}\right)\right]^2.$$

When $t(\cdot)$ is the square root transformation, the principal term of this expression becomes

$$\text{var}\left\{\sqrt{\hat{f}(y)}\right\} \approx \frac{1}{4}\frac{1}{nh}\alpha(w),$$

which does not depend on the unknown density f. This is the analogue for density estimation of the argument employed by Tukey (1977) in the derivation of the 'hanging rootogram' for histograms, also discussed by Scott (1992, Section 3.2.1).

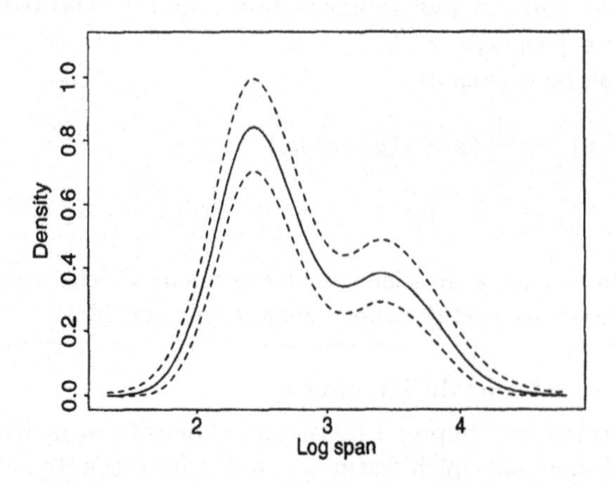

FIG. 2.2. A density estimate and variability bands from the aircraft span data.

On this square root scale the variability of the density estimate can now be expressed by placing a band of width two standard errors (se), $2\sqrt{\alpha(w)/4nh} = \sqrt{\alpha(w)/nh}$ around $\sqrt{\hat{f}}$. The edges of this band can then be translated back to the original scale. This has been carried out for the aircraft span data, for the third time period, in Fig. 2.2. For these particular data some caution needs to be exercised since the plausibility of the assumption of independent random sampling is in doubt. However, a band of this type requires careful interpretation under any circumstances. It is not a confidence band for the true density function because of the presence of bias, for which no adjustment has been made. In order to reflect the fact that only variance is being assessed, the term *variability band* will be used. This follows the terminology of Simonoff (1996) who described a similar display as a 'variability plot'.

Bands of this type are unlikely to provide a powerful means of examining particular hypotheses about a density function, particularly in view of their pointwise nature. However, the graphical display of variance structure can be helpful. For example, the width of the bands for the aircraft span data strengthens the evidence that the shoulder in the right hand tail of the estimate is a genuine feature, particularly since the effect of bias is to diminish modes and raise intervening troughs. The strength of any feature of this sort is therefore likely to be underestimated.

S-Plus Illustration 2.2. Variability bands for a density estimate

---

Figure 2.2 was constructed with the following S-Plus *code.*

```
provide.data(aircraft)
y <- log(Span)[Period==3]
sm.density(y, xlab = "Log span", display = "se")
```

2.4 Methods of choosing a smoothing parameter

In order to construct a density estimate from observed data it is necessary to choose a value for the smoothing parameter h. In this section the asymptotically optimal choice for h, and three of the most common, and arguably most effective, practical strategies, are described.

2.4.1 *Optimal smoothing*

An overall measure of the effectiveness of $\hat{f}$ in estimating f is provided by the mean integrated squared error described in Section 2.2. From the approximate expression given there it is straightforward to show that the value of h which minimizes MISE in an asymptotic sense is

$$h_{\text{opt}} = \left\{ \frac{\gamma(w)}{\beta(f)n} \right\}^{1/5}, \tag{2.4}$$

where $\gamma(w) = \alpha(w)/\sigma_w^4$, and $\beta(f) = \int f''(y)^2 dy$. This optimal value for h cannot immediately be used in practice since it involves the unknown density function f. However, it is very informative in showing how smoothing parameters should decrease with sample size, namely proportionately to $n^{-1/5}$, and in quantifying the effect of the curvature of f through the factor $\beta(f)$.

2.4.2 *Normal optimal smoothing*

Evaluation of the optimal formula for h when f is a normal density yields the simple formula

$$h = \left(\frac{4}{3n} \right)^{1/5} \sigma,$$

where σ denotes the standard deviation of the distribution. Clearly, the assumption of normality is potentially a self-defeating one when attempting to estimate a density nonparametrically but, for unimodal distributions at least, it gives a useful choice of smoothing parameter which requires very little calculation.

This approach to smoothing has the potential merit of being cautious and conservative. The normal is one of the smoothest possible distributions and so the optimal value of h will be large. If this is then applied to non-normal data it will tend to induce oversmoothing. The consequent reduction in variance at least has the merit of discouraging overinterpretation of features which may in fact be due to sampling variation. All of the one- and two-dimensional density estimates based on unrestricted continuous data in Chapter 1 were produced with normal optimal smoothing parameters.

In order to accommodate long tailed distributions and possible outliers, a robust estimate of σ is sometimes preferable to the usual sample standard deviation. A simple choice is the median absolute deviation estimator

$$\tilde{\sigma} = \text{median}\{|y_i - \tilde{\mu}|\}/0.6745,$$

where $\tilde{\mu}$ denotes the median of the sample; see Hogg (1979).

Normal optimal smoothing parameters can also be found in the multidimensional case. These are given by

$$h_i = \left\{ \frac{4}{(p+2)n} \right\}^{1/(p+4)} \sigma_i,$$

where p denotes the number of dimensions, h_i denotes the optimal smoothing parameter and σ_i the standard deviation in dimension i. For practical implementation the latter is replaced by a sample estimate. In the case of the aircraft data this simple form of smoothing is seen to be very effective in the two-dimensional displays of Section 1.3, even in the presence of a multimodal structure. The three-dimensional density contour produced from the geyser data, displayed in Fig. 1.9, was also constructed from normal optimal smoothing parameters.

This principle was carried to its logical extension by Terrell and Scott (1985) and Terrell (1990), who defined the 'oversmoothed' bandwidth to be the largest possible value of the optimal smoothing parameter, over all distributions with the same variance. This turns out to be fractionally larger than the normal optimal value.

2.4.3 *Cross-validation*

The ideas involved in cross-validation are given a general description by Stone (1974). In the context of density estimation, Rudemo (1982) and Bowman (1984) applied these ideas to the problem of bandwidth choice, through estimation of the integrated squared error (ISE)

$$\int \{\hat{f}(y) - f(y)\}^2 dy = \int \hat{f}(y)^2 dy - 2 \int f(y)\hat{f}(y) dy + \int f(y)^2 dy.$$

The last term on the right hand side does not involve h. The other terms can be estimated by

$$\frac{1}{n} \sum_{i=1}^{n} \int \hat{f}_{-i}^2(y) dy - \frac{2}{n} \sum_{i=1}^{n} \hat{f}_{-i}(y_i), \qquad (2.5)$$

where $\hat{f}_{-i}(y)$ denotes the estimator constructed from the data without the observation y_i. It is straightforward to show that the expectation of this expression is the MISE of $\hat{f}$ based on $n-1$ observations, omitting the $\int f^2$ term. The value of h which minimises this expression therefore provides an estimate of the optimal smoothing parameter. Stone (1984) derived an asymptotic optimality result for bandwidths which are chosen in this cross-validatory fashion. From a computational point of view, the use of normal kernel functions allows the integrals in (2.5) to be evaluated easily, by expanding $\hat{f}_{-i}^2(x)$ into its constituent terms and applying results on the convolution of normal densities.

The left panel of Fig. 2.3 displays the cross-validation function (2.5) for the tephra data. This shows a minimum at $h = 0.018$. The right panel of the figure shows the density estimate produced by this value of smoothing parameter.

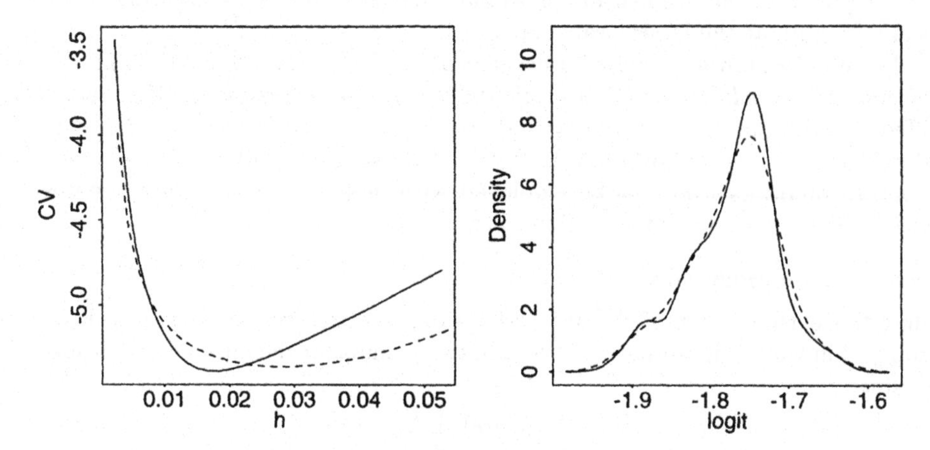

FIG. 2.3. The left panel shows the cross-validation function based on the tephra data, on a logit scale. The MISE function for a normal density has been superimposed, on a shifted vertical scale, for comparison. The right panel shows density estimates from the tephra data using the cross-validatory smoothing parameter (full line), and the normal optimal smoothing parameter (dashed line).

Cross-validation has therefore successfully chosen a suitable amount of smoothing to apply to the data. In the left panel, the MISE function for a normal distribution is also displayed. Since the vertical position is unimportant the function has been shifted to allow it to be superimposed on the graph. The normal optimal smoothing parameter in this case has the value 0.026. Cross-validation has attempted to take account of the mild skewness in the data. The smaller smoothing parameter helps to accentuate the peak of the density, but has the additional effect of producing a less smooth left hand tail.

Of the estimates produced so far from the tephra data, the variable bandwidth approach displayed in Fig. 2.1 seems the most effective choice for these data. It is an advantage of the cross-validatory approach that its general definition allows it to be applied in a wide range of settings. It can, for example, be applied to the choice of the overall smoothing parameter h in the variable bandwidth form $h_i = hd_k(y_i)$. The general computational form of the cross-validation function, using normal kernels, is given at the end of this section. Minimisation of this for a kernel estimator with normalised nearest neighbour weights ($k = 7$) on a square root scale produces a smoothing parameter of 0.022 for the tephra data. This is very close to the normal optimal value and so the resulting estimate is very similar to that displayed in Fig. 2.1.

Local minima can occur in cross-validation functions, as described by Wand and Jones (1995, Section 3.3), and so it can be wise to employ plotting, in addition to a numerical algorithm, to locate the minimising smoothing parameter.

Marron (1993) recommends using the smoothing parameter corresponding to the local minimum at the largest value of h.

Techniques known as *biased cross-validation* (Scott and Terrell 1987) and *smoothed cross-validation* (Hall *et al.* 1992) also aim to minimise ISE but use different estimates of this quantity to do so. These approaches are also strongly related to the 'plug-in' approach described in the following subsection.

Cross-validation can also be employed with multivariate density estimates since (2.5) remains a valid definition in several dimensions.

2.4.4 *Plug-in bandwidths*

Since the earliest days of density estimation, iterative procedures have been proposed in which an estimate $\hat{f}$ is used in the formula for the optimal smoothing parameter:

$$h = \left\{ \frac{\gamma(w)}{\beta(\hat{f})n} \right\}^{1/5}.$$

Scott *et al.* (1977) is an early example of this approach. If normal kernels are used, $\gamma(w)$ and $\beta(\hat{f})$ can be calculated relatively easily and the value of h which solves this equation can be found by a suitable numerical algorithm.

Within the past few years, considerable progress has been made in the development of this approach. In particular, Sheather and Jones (1991), extending the work of Park and Marron (1990), described a bandwidth selection procedure with excellent properties. This is based on a clever method of estimation of f'', using an additional smoothing parameter related to h for this purpose, guided by asymptotic theory on the estimation of derivatives of a density, which requires a larger amount of smoothing than in the estimation of f itself.

This estimator has very good finite sample, as well as asymptotic, properties. In particular, it is more stable than the cross-validation approach described above. The two techniques take different approaches to the same problem of minimising ISE. Cross-validation estimates the ISE function and locates the minimum. The plug-in approach minimises the function theoretically and then estimates this minimising value directly. The good performance of this estimation process produces a method which is subject to less variability.

With the tephra data, the Sheather–Jones method produces a smoothing parameter of 0.016. This is very close to the cross-validatory value of 0.018 and so in this particular example there is little difference between the two approaches.

2.4.5 *Discussion*

How best to choose smoothing parameters is still the subject of much research and debate. For one-dimensional data the 'plug-in' approach of Sheather and Jones (1991) seems to be very effective. In contrast, there is both empirical and theoretical evidence that cross-validation is subject to greater variability. However, it does have the advantage of providing a general approach which can be adapted to a wide variety of settings, including multivariate data and variable

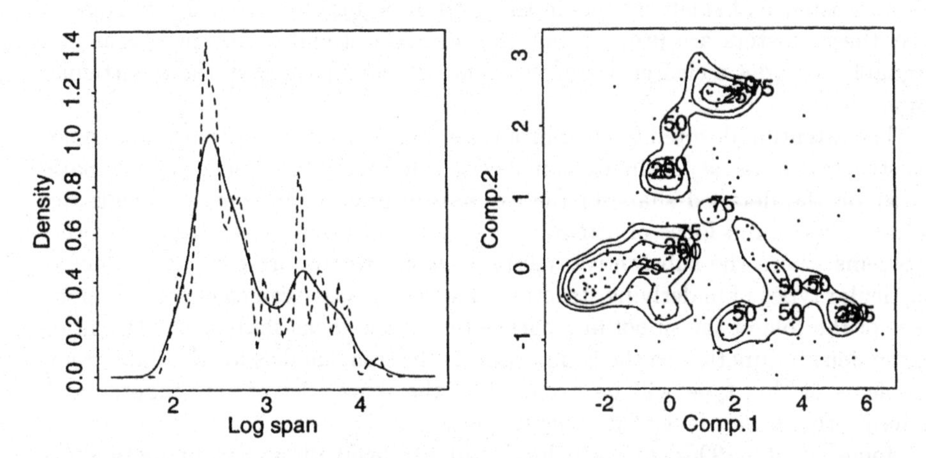

FIG. 2.4. The left panel displays density estimates produced from the log span data, using the Sheather–Jones plug-in (full line) and cross-validatory (dashed line) smoothing parameters. The right panel displays a density contour plot of the first two principal components of the aircraft data, using cross-validation to select the parameters in two dimensions.

bandwidths. Other examples of its versatility are provided by Fisher *et al.* (1987), who describe a method for spherical data, and Bowman and Prvan (1996), who show how the method can be applied to the smoothing of distribution functions.

As a further illustration, the left panel of Fig. 2.4 displays density estimates produced from the log span data, considered extensively in Chapter 1. The full line corresponds to the use of the Sheather–Jones smoothing parameter. This seems an effective choice, which captures the structure of the data, with a shoulder in the right hand tail. The dashed line corresponds to use of the cross-validatory smoothing parameter. This is very small and results in a highly variable estimate. Silverman (1986, Section 3.4.3) gives an explanation of the fact that cross-validation is rather sensitive to the presence of clustering in the data. There are several repeated values in the aircraft span data and this results in a very small smoothing parameter, as the method attempts to reproduce this granularity in the density estimate.

In two dimensions, the smoothing parameters (h_1, h_2) can be chosen to minimise the cross-validatory criterion (2.5) when $\hat{f}$ is a bivariate estimate. To reduce the complexity of this problem it is convenient to use a simple scaling for each dimension, given by $h_1 = a\hat{\sigma}_1$, $h_2 = a\hat{\sigma}_2$, and to minimise this over a. This has been done in the right hand panel of Fig. 2.4, where the contours of a density estimate, using the aircraft data from the third time period, and using cross-validatory smoothing parameters, are displayed. This plot should be contrasted with Fig. 1.7, which displays a density estimate for the same data using a normal optimal smoothing parameter. Again, cross-validation has reacted strongly to the existence of small clusters in the data. It seems clear from the scatterplot that these clusters are present, but the normal optimal smoothing parameter arguably provides a clearer overall summary of the structure in these particular data.

The attention devoted to an appropriate choice of smoothing parameter when constructing a density estimate is motivated by the large influence of this parameter on the detailed shape of the estimate produced. In some problems this is indeed the main focus of attention. There is, however, quite a range of other problems where the substantive questions relate to the presence or absence of particular types of underlying structure. This perspective can change the importance of the role of the smoothing parameter, as the focus shifts from estimation of the density function to the comparison of different models for the data. Some problems of this type will be explored in the remainder of this chapter and, indeed, are a major focus throughout the book.

Jones *et al.* (1996) give a helpful and balanced discussion of methods of choosing the smoothing parameter in density estimation.

S-Plus Illustration 2.3. Data based choices of smoothing for the tephra data

---

Figure 2.3 was constructed with the following S-Plus *code. The function* `nmise` *calculates the MISE for a density estimate of a normal distribution. It is supplied with the* `sm` *library.*

```
provide.data(tephra)
logit <- log(Al2O3/(100-Al2O3))
par(mfrow=c(1,2))
h.cv <- hcv(logit, display = "lines", ngrid = 32)
n  <- length(logit)
sd <- sqrt(var(logit))
h  <- seq(0.003, 0.054, length=32)
lines(h, nmise(sd, n, h) - 5.5, lty = 3)
sm.density(logit, h.cv)
sm.density(logit, lty = 3, add = T)
par(mfrow=c(1,1))
```

S-Plus Illustration 2.4. Data based choices of smoothing for the aircraft data

---

Figure 2.4 was constructed with the following S-Plus *code.*

```
provide.data(aircraft)
provide.data(airpc)
y <- log(Span)[Period==3]
par(mfrow=c(1,2))
sm.density(y, h = hcv(y), xlab="Log span", lty=3, yht=1.4)
sm.density(y, h = hsj(y), add = T)
```

```
pc3 <- cbind(Comp.1, Comp.2)[Period==3,]
sm.density(pc3, h = hcv(pc3), display = "slice")
par(mfrow=c(1,1))
```

Mathematical aspects: Convolutions of normal densities

A variety of calculations with density estimates constructed from normal kernels lead to expressions involving the convolution of two normal densities. The following derivation of the computational form of the cross-validatory criterion is one example (it is convenient to give this convolution result a separate statement, for future reference):

$$\int \phi(y-\mu_1;\sigma_1)\phi(y-\mu_2;\sigma_2)dy = \phi\left(\mu_1-\mu_2;\sqrt{\sigma_1^2+\sigma_2^2}\right).$$

Here $\phi(z;\sigma)$ denotes the normal density function with mean 0 and standard deviation σ.

Mathematical aspects: The computational form of the cross-validatory criterion

The cross-validatory criterion (2.5) can be written in a more computationally convenient form when normal kernels are used. To derive a general expression, variable bandwidths h_i will be used. In the one-dimensional case, the criterion becomes

$$\begin{aligned}
&\frac{1}{n}\sum_{i=1}^{n}\int \hat{f}_{-i}^2(y)dy - \frac{2}{n}\sum_{i=1}^{n}\hat{f}_{-i}(y_i) \\
&= \frac{1}{n}\sum_{i=1}^{n}\int \frac{1}{(n-1)^2}\sum_{j\neq i}\sum_{k\neq i}\phi(y-y_j;h_j)\phi(y-y_k;h_k)dy \\
&\qquad -\frac{2}{n(n-1)}\sum_{i\neq j}\phi(y_i-y_j;h_j) \\
&= \frac{1}{n(n-1)}\sum_{i=1}^{n}\phi(0;\sqrt{2}h_i) + \frac{(n-2)}{n(n-1)^2}\sum_{i\neq j}\phi\left(y_i-y_j;\sqrt{h_i^2+h_j^2}\right) \\
&\qquad -\frac{2}{n(n-1)}\sum_{i\neq j}\phi(y_i-y_j;h_j).
\end{aligned}$$

In the two-dimensional case, with product kernels $\phi(y_1-y_{1i};h_{1i})\phi(y_2-y_{2i};h_{2i})$, this becomes

$$\begin{aligned}
&\frac{1}{n(n-1)}\sum_{i=1}^{n}\phi(0;\sqrt{2}h_{1i})\phi(0;\sqrt{2}h_{2i}) \\
&+\frac{(n-2)}{n(n-1)^2}\sum_{i\neq j}\phi\left(y_{1i}-y_{1j};\sqrt{h_{1i}^2+h_{1j}^2}\right)\phi\left(y_{2i}-y_{2j};\sqrt{h_{2i}^2+h_{2j}^2}\right)
\end{aligned}$$

$$-\frac{2}{n(n-1)}\sum_{i\neq j}\phi(y_{1i}-y_{1j};h_{1j})\phi(y_{2i}-y_{2j};h_{2j}).$$

2.5 Testing normality

The tephra data exhibit some negative skewness. Before settling on a model for these data it would be helpful to assess whether this skewness is a real feature or simply the result of random variation. A probability plot of the data is a standard graphical device for checking normality. The left panel of Fig. 2.5 shows such a plot, from which it remains unclear whether there is strong evidence of non-normality. There are very many formal tests of normality; Stephens (1974) provides a comprehensive review. A simple but powerful approach is to assess the straightness of a probability plot, as described for example by Shapiro and Wilk (1965) and Filliben (1975). Applied to the tephra data this latter test does not detect significant non-normality.

One use of a nonparametric approach to modelling is in assessing the fit of a parametric model by comparing the two. The existence of density estimates raises the possibility that effective comparisons might be made on the natural density scale, rather than from a probability plot or empirical distribution function, which many other tests use. Goodness-of-fit statistics based on density functions have been proposed by Bickel and Rosenblatt (1973) and Bowman (1992) among others. The addition of new tests in this much studied area needs to be well motivated. However, density based statistics are of interest because they penalise departures from the hypothesised distribution in a different, and arguably more intuitive, way than more traditional procedures. They also have the potential to be equally easy to implement for multivariate data, which is not true of many other techniques.

There is a very large number of possible ways in which a proposed and an estimated density can be compared in a test statistic. One possibility is the integrated squared error

$$\int\left\{\hat{f}(y)-\phi\left(y-\hat{\mu};\sqrt{\hat{\sigma}^2+h^2}\right)\right\}^2 dy.$$

Notice that the density estimate is not compared to the proposed normal density $\phi(y-\hat{\mu};\hat{\sigma})$. Since the smoothing operation produces bias, it is more appropriate to contrast the density estimate with its mean under the assumption of normality, which is shown at the end of this section to be $\phi(y-\mu;\sqrt{\sigma^2+h^2})$.

In order to remove the effects of location and scale the data can be standardised to have sample mean 0 and sample variance 1. The test statistic on standardised data then becomes

$$\int\left\{\hat{f}(y)-\phi\left(y;\sqrt{1+h^2}\right)\right\}^2 dy.$$

When the kernel function is a normal density a convenient computational form of the statistic can be derived from results on convolutions. The statistic has a

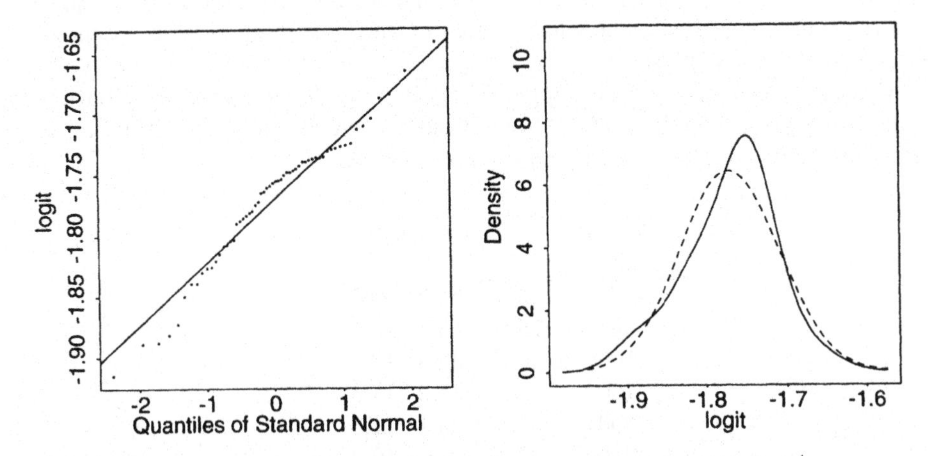

FIG. 2.5. The left panel shows a probability plot of the tephra data. The right panel shows a density estimate (full line) of the tephra data, and the mean value (dashed line) of this estimate under the assumption that the data are normally distributed.

mathematical structure which is similar to many of the well known test statistics for normality, such as those based on empirical distribution functions. However, as is the case with many goodness-of-fit statistics, it is computationally straightforward to derive percentage points by simulation.

In order to carry out the test, it is necessary to decide on the value of the smoothing parameter. Since this is a test for normality, it is natural to use normal optimal smoothing. When normality is an appropriate assumption the smoothing parameter will also be appropriate. When the data are not normally distributed, this form of smoothing will induce conservative behaviour, since the large value of the smoothing parameter will not promote the display of spurious features.

When applied to the tephra data, where the sample size is 59, the test statistic takes the value 0.00908. This is greater than the critical value in the table overleaf (divided by 1000), which was produced by simulation. The test has therefore identified significant non-normality. The different nature of the test statistic has allowed it to detect features which the probability plot did not.

Bowman (1992) investigated the power of the ISE statistic and found it to perform well in comparison with the best of the traditional procedures, such as the Shapiro–Wilk and Anderson–Darling tests. However, the strongest argument for the study of such tests is the ease with which analogous procedures can be defined in more than one dimension. Bowman and Foster (1993) investigated this for the multivariate normal distribution and provided tables of critical values. With multivariate data it will often be informative to construct Mahalanobis probability plots, or other diagnostic procedures which focus on important kinds of departures from normality. However, there is also a role for global tests of the kind described above in cases such as discrimination, and exploratory techniques such as projection pursuit, where local values of the density function are essential components of the analysis.

The table below records the upper 5% points of the simulated distribution of the ISE statistic (multiplied by 1000 for convenience) under normality, for a variety of sample sizes and for up to three dimensions.

Sample	Dimensions		
size	1	2	3
25	109.0	64.2	30.3
50	76.6	50.7	23.1
100	56.7	35.5	16.7
150	45.3	27.2	14.3
200	38.0	23.1	11.8
250	33.2	20.5	10.9
300	30.1	18.9	9.64
350	27.2	16.9	8.89
400	23.2	16.1	8.50
500	20.5	13.9	7.47

S-Plus Illustration 2.5. Assessing normality for the tephra data

Figure 2.5 was constructed with the following S-Plus *code.*

```
provide.data(tephra)
logit <- log(Al2O3/(100-Al2O3))
par(mfrow=c(1,2))
qqnorm(logit)
qqline(logit)
cat("ISE statistic:", nise(logit),"\n")
sm <- sm.density(logit)
y  <- sm$eval.points
sd <- sqrt(hnorm(logit)^2 + var(logit))
lines(y, dnorm(y, mean(logit), sd), lty = 3)
par(mfrow=c(1,1))
```

Mathematical aspects: The exact mean and variance of a density estimate constructed from normal data

Using the result on convolutions of normal densities given at the end of Section 2.4, it follows that when the data have a normal distribution with mean μ and standard deviation σ, the mean of the density estimate is

$$\mathbb{E}\left\{\hat{f}(y)\right\} = \mathbb{E}_z\left\{\phi(y - z; h)\right\}$$

$$= \int \phi(y-z;h)\phi(z-\mu;\sigma)dz$$
$$= \phi\left(y-\mu;\sqrt{h^2+\sigma^2}\right).$$

For completeness, an expression is also derived for the variance:

$$\begin{aligned}
\mathrm{var}\left\{\hat{f}(y)\right\} &= \frac{1}{n}\mathrm{var}_z\{\phi(y-z;h)\} \\
&= \frac{1}{n}[\mathbb{E}_z\{\phi^2(y-z;h)\} - \mathbb{E}_z\{\phi(y-z;h)\}^2] \\
&= \frac{1}{n}\left[\phi\left(0;\sqrt{2}\,h\right)\phi\left(y-\mu;\sqrt{\sigma^2+\tfrac{1}{2}h^2}\right) - \phi\left(y-\mu;\sqrt{\sigma^2+h^2}\right)^2\right].
\end{aligned}$$

This will be used in Section 2.6.

2.6 Normal reference band

In the discussion of the test of normality it was natural to consider a graphical representation of the density estimate and a normal density curve, as in Fig. 2.5. This idea can be extended further. It was shown at the end of the previous section that when the true density function is normal with mean μ and variance σ^2, and the kernel function w is also normal, the mean and variance of the density estimate at the point y are

$$\begin{aligned}
\mathbb{E}\left\{\hat{f}(y)\right\} &= \phi\left(y-\mu;\sqrt{h^2+\sigma^2}\right), \\
\mathrm{var}\left\{\hat{f}(y)\right\} &= \frac{1}{n}\left[\phi\left(0;\sqrt{2}\,h\right)\phi\left(y-\mu;\sqrt{\sigma^2+\tfrac{1}{2}h^2}\right) - \phi\left(y-\mu;\sqrt{\sigma^2+h^2}\right)^2\right].
\end{aligned}$$

These expressions allow the likely range of values of the density estimate to be calculated, under the assumption that the data are normally distributed. This can be expressed graphically through a *reference band.* At each point y of interest, the band is centred at $\mathbb{E}\left\{\hat{f}(y)\right\}$ and extends a distance $2se\{\hat{f}(y)\}$ above and below. The sample mean and variance can be used in place of the normal parameters μ and σ^2.

Figure 2.6 has reference bands superimposed on density estimates for the tephra data, using two different smoothing parameters. In both cases the peak of the observed curve lies just outside the reference band. As a follow-up to the goodness-of-fit test, this indicates a sharper peak than a normal density and the presence of some negative skewness. It should be emphasised that this is not intended as a replacement for the global test of normality. It does, however, provide a useful graphical follow-up procedure by indicating the likely position of a density estimate when the data are normal. This can be helpful in identifying where non-normal features might lie, or in understanding, through the displayed standard error structure, why some apparently non-normal feature has not led to a significant test result.

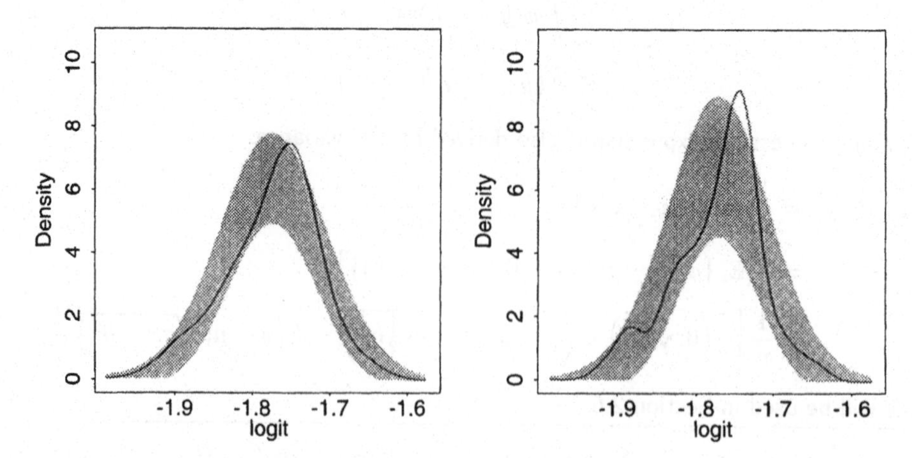

FIG. 2.6. Density estimates of the tephra data, with reference bands for a normal distribution. The left panel displays an estimate and band using the normal optimal smoothing parameter. The right panel uses the Sheather–Jones plug-in smoothing parameter.

S-Plus Illustration 2.6. Reference bands for normality with the tephra data

---

Figure 2.6 was constructed with the following S-Plus *code.*

```
provide.data(tephra)
logit <- log(Al2O3/(100-Al2O3))
par(mfrow=c(1,2))
sm.density(logit, model = "Normal")
sm.density(logit, h = hsj(logit), model = "Normal")
par(mfrow=c(1,1))
```

2.7 Testing independence

In two dimensions, other statistical problems become important. One is to assess the presence, or strength, of dependence between two variables. While the correlation coefficient is very useful for linear structures, there are nonlinear patterns which can be missed. An example arises with the aircraft data. Figure 2.7 displays a scatterplot of the variables span and speed, both on log scales, for the third time period. In order to assess whether these two variables are associated, a natural starting point is the correlation coefficient. For these data this takes the value 0.017, which is so close to 0 that it does not provide even a hint of linear correlation. A standard nonparametric approach is to calculate the rank correlation coefficient. This takes the value 0.023, which again gives no evidence of a relationship between the two variables.

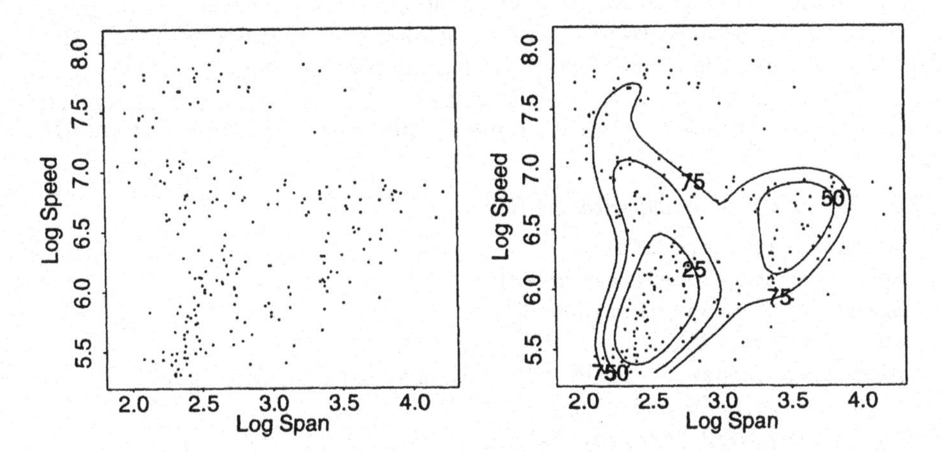

FIG. 2.7. The left panel displays the aircraft span and speed data, on log scales, for the third time period, 1956–1984. The right panel shows the same data with a density contour plot superimposed.

The existence of density estimates gives the opportunity to explore dependence in a more flexible way. A general definition of independence is that the joint density $f_{12}(y_1, y_2)$ of two variables decomposes into the product of the marginal densities $f_1(y_1)$ and $f_2(y_2)$. An assessment of independence in a very general sense can then be performed by contrasting these joint and marginal density estimates. There are many ways in which this could be done. However, a natural approach is to construct a likelihood ratio expression

$$\frac{1}{n}\sum \log\left\{\frac{\hat{f}_{12}(y_{1i}, y_{2i})}{\hat{f}_1(y_{1i})\hat{f}_2(y_{2i})}\right\}.$$

Kent (1983) and Joe (1989) considered statistics of this kind. A computational approach to the derivation of the distribution of this test statistic under the null hypothesis of independence is to apply a permutation argument, where values of y_{1i} are associated with randomly permuted values of y_{2i}. An empirical p-value can be calculated as the proportion of statistics computed from the permuted data whose values exceed that of the observed statistic from the original data. With the aircraft span and speed data, the empirical p-value is 0.00, based on a simulation size of 200.

The striking difference in the linear and nonparametric assessments of association between these variables is explained by the right hand panel of Fig. 2.7. The density contours show clear association, with span and speed generally increasing together, but with a significant cluster of observations falling in the region of low span and high speed. The presence of these observations causes the linear association between the variables to reduce almost to 0.

Bjerve and Doksum (1993), Doksum *et al.* (1994) and Jones (1996) suggest how dependence between variables can be quantified in a local way through the definition of correlation curves and local dependence functions.

S-Plus Illustration 2.7. Density contour plots for exploring independence

---

Figure 2.7 was constructed with the following S-Plus *code.*

```
provide.data(aircraft)
Speed3 <- log(Speed[Period==3])
Span3  <- log(Span[Period==3])
par(mfrow=c(1,2))
plot(Span3, Speed3, xlab = "Log Span", ylab = "Log Speed")
air3  <- cbind(Span3, Speed3)
sm.density(air3, display="slice",
        xlab= "Log Span", ylab = "Log Speed")
par(mfrow=c(1,1))
```

An additional script to carry out the permutation test is available.

2.8 The bootstrap and density estimation

The material of earlier sections shows that the form of the bias and variance of a density estimate complicates any attempt to carry out simple forms of inference, including even a simple confidence interval for the density estimate at a point. In such situations, the bootstrap sometimes provides an attractive tool, giving a potential means of carrying out inference by resampling methods, where analytic methods prove too complex. Davison and Hinkley (1997) give a general description of the bootstrap method.

A simple bootstrap procedure for density estimates is as follows.

1. Construct a density estimate $\hat{f}$ from the observed data $\{y_1, \ldots, y_n\}$.
2. Resample the data $\{y_1, \ldots, y_n\}$ with replacement to produce a bootstrap sample $\{y_1^*, \ldots, y_n^*\}$.
3. Construct a density estimate $\hat{f}^*$ from the bootstrap data $\{y_1^*, \ldots, y_n^*\}$.
4. Repeat steps 2 and 3 a large number of times to create a collection of bootstrap density estimates $\{\hat{f}_1^*, \ldots, \hat{f}_B^*\}$.
5. Use the distribution of $\hat{f}_i^*$ about $\hat{f}$ to mimic the distribution of $\hat{f}$ about f.

It is worthwhile checking the validity of step 5. To do this, consider the mean of $\hat{f}^*$. Since the distribution of y_i^* is uniform over $\{y_1, \ldots, y_n\}$, it follows that

$$\mathbb{E}\left\{ \hat{f}^*(y) \right\} = \mathbb{E}\{w(y - y_i^*; h)\} = \hat{f}(y),$$

from which it is immediately apparent that the bias which is present in the distribution of $\hat{f}$ is absent in the bootstrap version. The bootstrap distribution of $\hat{f}^*$ is therefore missing an essential component. Hall (1992) gives appropriate

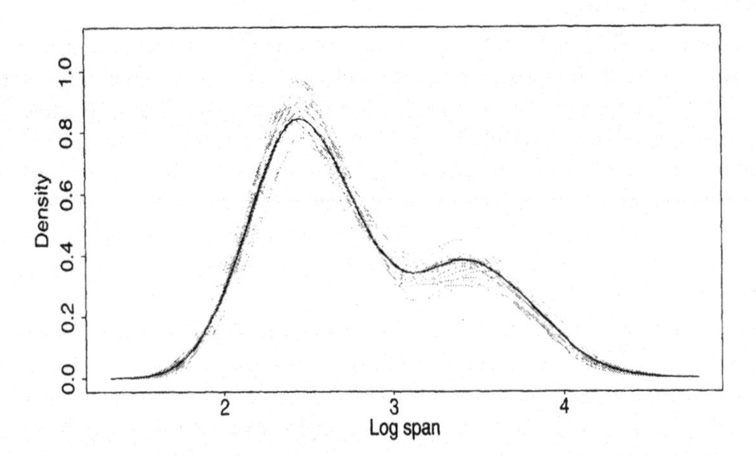

FIG. 2.8. Bootstrap density estimates from the aircraft span data, as an alternative method of constructing variability bands.

theory for this situation and shows that the bootstrap does correctly mimic the variance of $\hat{f}$. Bootstrapping can therefore be used to provide an alternative means of generating the variability bands described in Section 2.3. Figure 2.8 displays density estimates from multiple bootstrap samples of the aircraft span data. This is a bootstrap analogue of Fig. 2.2.

Bootstrapping has a very useful role to play and it can be particularly helpful in some hypothesis testing contexts. Silverman (1981) describes an interesting case where inference can be carried out on the number of modes present in a density, using a technique known as the smoothed bootstrap, which involves simulating from $\hat{f}$ rather than resampling the original data. Taylor (1989) used the smoothed bootstrap to construct an estimate of mean integrated squared error and proposed this as a means of selecting a suitable smoothing parameter. Scott (1992, Section 9.3.2) discusses and illustrates the role of the smoothed bootstrap in constructing confidence intervals.

S-Plus Illustration 2.8. Bootstrapping density estimates

Figure 2.8 was constructed with the following S-Plus *code.*

```
provide.data(aircraft)
y <- log(Span)[Period==3]
sm.density(y, xlab = "Log span")
for (i in 1:20) sm.density(sample(y, replace=T), col=6, add=T)
sm.density(y, xlab = "Log span", add=T)
```

The number of bootstrap estimates produced is controlled by the range of `i` *in the* `for` *loop. On some computers these simulations may take a long time.*

2.9 Further reading

Wand and Jones (1995) provide a very good overview of current research in density estimation. The chapter on bandwidth choice gives a particularly thorough discussion of this important topic, and more technical aspects of density estimation are covered in some detail. Scott (1992) gives a rounded discussion of the multivariate case, while Fan and Gijbels (1996) review density estimation within a broader context of smoothing techniques, including a description of the wavelet approach. Devroye and Györfi (1985) take an intriguingly different approach to the whole area by considering performance in terms of the L_1 norm rather than mean integrated squared error.

It was stated earlier that the detailed shape of the kernel function is relatively unimportant. It is, however, possible to derive an asymptotically optimal kernel shape. This was done by Epanechnikov (1969) who showed this to be quadratic.

Kernel shape has also been proposed as a means of ameliorating the awkward reality of bias in density estimates, since a suitable choice of kernel function can remove further terms in the bias expansion. This necessitates that the kernel function applies negative weight in some sections of the axis. Granovsky and Müller (1991) describe research in this area, while Marron and Wand (1992) report a study of practical performance.

Hall and Marron (1988) and Hall (1990) are examples of papers which provide asymptotic theory for variable bandwidth estimators.

For tests of normality, the nature of a density function allows approaches which are not possible on other scales. For example, Vasicek (1976) proposed a test statistic based on entropy, $-\int f(y)\log f(y)dy$, a quantity which is maximised, over all distributions with fixed variance, at the normal distribution. Bowman and Foster (1993) used a kernel approach for $\hat{f}$ to investigate the power of a sample version, $-\frac{1}{n}\sum_{i=1}^{n}\log \hat{f}(y_i)$.

Exercises

2.1 *Variability bands.* Construct variability bands for the aircraft span data, as in Fig. 2.8, and superimpose variability bands derived from the square root variance stabilising argument, in order to compare the two. Repeat this with the tephra data.

2.2 *Comparing methods of choosing smoothing parameters.* Simulate data from a distribution of your choice and compare the density estimates produced by smoothing parameters which are chosen by the normal optimal, Sheather–Jones and cross-validatory approaches. (These are available in the functions `hnorm`, `hsj` and `hcv`.) In each case, try to make your own judgement on a good value of smoothing parameter and compare this with the one chosen by the automatic methods. Where there are differences, see if you can identify the cause.

Examine in particular the variability of the smoothing parameters selected by the Sheather–Jones method and by cross-validation.

Repeat this with data simulated from other distributions.

2.3 *Cross-validation with nearest neighbour distances.* Use the `hweights` argument of `hcv` to find the cross-validatory smoothing parameter for a particular value of k with the tephra data. Repeat for different values of k and so find the cross-validatory solution over h and k simultaneously.

2.4 *Cross-validation and clustered data.* The S-Plus function `jitter` randomly perturbs data by a small amount. Apply this function to the aircraft span data for the third time period to see whether this reduces the problem caused by repeated values when choosing a smoothing parameter by cross-validation.

2.5 *Assessing normality for the tephra data.* Investigate whether the test of normality on the tephra data identifies non-normality on the original, rather than logit, scale.
The function `nise`, supplied in the `sm` library, computes the test statistics for normality described in Section 2.5. Write S-Plus code which will compute an empirical p-value for the test by evaluating the statistic on simulated normal data. Apply this test to the tephra data using a variety of different smoothing parameters and examine the extent to which the conclusions of the test are affected by the particular bandwidth used.
Normal reference bands for the tephra data are shown in Fig. 2.6. Find the range of values of the smoothing parameter h for which the density estimate lies outside these bands.

2.6 *Aircraft data.* Apply the test of normality to the aircraft span data for the third time period, and construct an associated reference band, as a simple means of investigating whether the asymmetry in the density estimate is more than would be expected from normal random variation. Explore the effect of changing the smoothing parameter on the relationship between the reference bands and the density estimate.
Repeat this for the other variables in the dataset.

2.7 *Data based smoothing parameters in a test of normality.* The test described in Section 2.5 used a normal optimal smoothing parameter whose value, on standardised data, depends only on sample size. Conduct a small simulation study to examine the effects of using a data based choice of smoothing parameter, such as the one produced by the Sheather–Jones method, on the size and power of the test.

2.8 *Mean and variance of a smooth estimate of a distribution function.* A kernel approach to constructing a smooth estimate $\hat{F}(y)$ of a distribution function was described in Section 1.8. Use a Taylor series approach, analogous to the one described in Section 2.2, to derive approximate expressions for the mean and variance of $\hat{F}(y)$. Check your results against those of Azzalini (1981).

3

NONPARAMETRIC REGRESSION FOR EXPLORING DATA

3.1 Introduction

Regression is one of the most widely used of all statistical tools. Linear modelling in its widest sense is both well developed and well understood, and there is in addition a variety of useful techniques for checking the assumptions involved. However, there are cases where such models cannot be applied because of intrinsic nonlinearity in the data. Nonparametric regression aims to provide a means of modelling such data. Even where the suitability of linear models has not yet been brought into question, smoothing techniques are still useful by enhancing scatterplots to display the underlying structure of the data, without reference to a parametric model. This can in turn lead to further useful suggestions of, or checks on, appropriate parametric models. In this chapter, the main ideas of nonparametric regression are introduced and illustrated on several different types of data. The emphasis throughout is on the exploration of data, although technical and inferential issues will be raised for discussion in later chapters.

3.2 Basic ideas

The process of radiocarbon dating is widely used to identify the ages of archaeological objects. In order to calibrate the process, it is necessary to use samples of known age and to relate the 'radiocarbon ages' produced by the radiometric process to the true 'calendar ages', which are available from such sources as tree ring information. Figure 3.1 displays a subset of data published by Pearson and Qua (1993) which covers a large range of calendar ages. The left panel of this figure displays the data over the entire range of calendar ages, with the line $y = x$ superimposed as a reference. This shows that the radiocarbon ages do need to be adjusted to produce an estimate of true calendar age. It also shows that a simple linear adjustment will not suffice.

The right panel of Fig. 3.1 displays the data for a small section of the curve, corresponding to calendar ages between 2000 and 3000 years. A fitted linear regression has been superimposed to show that even on a small scale a linear model fails to capture the pattern in the regression function. It is well known that fluctuations in the natural production of radiocarbon can produce nonlinear effects in the calibration curve. A suitable model for these data is therefore

$$y = m(x) + \varepsilon, \tag{3.1}$$

where y denotes the response variable, x the covariate and ε denotes an independent error term with mean 0 and variance σ^2.

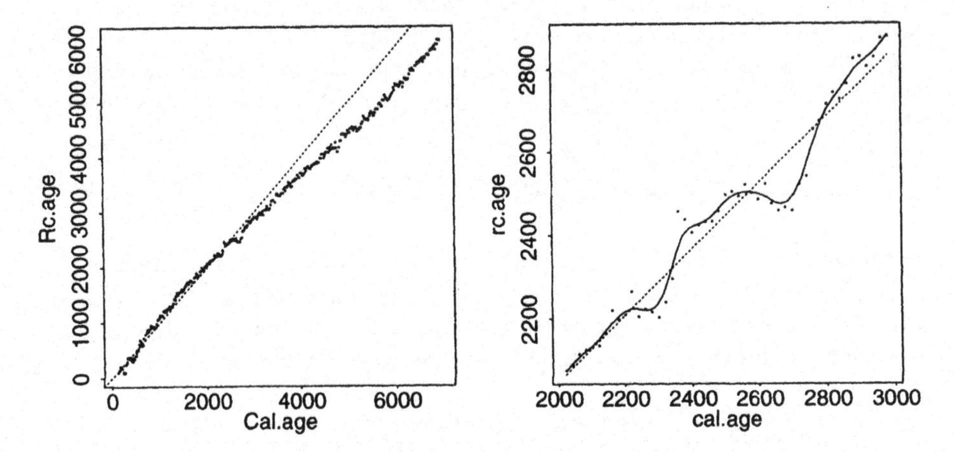

FIG. 3.1. Radiocarbon dating calibration data. In the left panel the curve $y = x$ is superimposed. In the right panel a nonparametric regression curve and a fitted linear regression have been superimposed.

Figure 3.1 also displays a smooth curve which estimates the underlying curve $m(x)$ and so offers a flexible description of the calibration relationship. Figure 3.2 indicates how a smooth curve of this type can be produced. Just as density estimation seeks to average counts of the data locally, so nonparametric regression seeks to average the values of the response variable locally. A simple kernel approach is to construct the *local mean* estimator

$$\tilde{m}(x) = \frac{\sum_{i=1}^{n} w(x_i - x; h) y_i}{\sum_{i=1}^{n} w(x_i - x; h)},$$

which was first proposed by Nadaraya (1964b) and Watson (1964). The kernel function $w(z; h)$ is generally a smooth positive function which peaks at 0 and decreases monotonically as z increases in size. This ensures that most weight is given to the observations whose covariate values x_i lie close to the point of interest x. For convenience, a normal density function, with standard deviation h, is commonly used as the kernel.

The *smoothing parameter* h controls the width of the kernel function, and hence the degree of smoothing applied to the data. This is illustrated in Fig. 3.2, in which the shaded regions represent the 'windows' in the x axis where the kernel places appreciable weight in constructing the local average, for a variety of values of smoothing parameter. When h is the standard deviation of a normal density, observations over an effective range of $4h$ in the covariate axis will contribute to the estimate. As the smoothing parameter increases, the resulting estimator misses some details in the curvature of the data. As the smoothing parameter decreases, the estimator begins to track the data too closely and will end up interpolating the observed points. Clearly, some effective compromise is required. This topic will be discussed in more detail in Chapter 4.

An alternative approach to the construction of a local mean for the data is to fit a *local linear regression*. This involves solving the least squares problem

$$\min_{\alpha,\beta} \sum_{i=1}^{n} \{y_i - \alpha - \beta(x_i - x)\}^2 \, w(x_i - x\,; h) \tag{3.2}$$

and taking as the estimate at x the value of $\hat{\alpha}$, as this defines the position of the local regression line at the point x. Again, it is the role of the kernel weights to ensure that it is observations close to x which have most weight in determining the estimate. The local linear estimator can be given an explicit formula,

$$\hat{m}(x) = \frac{1}{n} \sum_{i=1}^{n} \frac{\{s_2(x;h) - s_1(x;h)(x_i - x)\}\, w(x_i - x;h)\, y_i}{s_2(x;h)s_0(x;h) - s_1(x;h)^2}, \tag{3.3}$$

where $s_r(x;h) = \{\sum (x_i - x)^r w(x_i - x;h)\}/n$. The local mean estimator described above can be derived in a similar way, by removing the $\beta(x_i - x)$ term from the formulation of the least squares problem.

The idea of local linear regression was proposed some time ago by Cleveland (1979) and other authors. However, the idea has recently been the focus of considerable renewed interest. Fan and Gijbels (1992) and Fan (1993) showed the excellent theoretical properties which this estimator possesses. In particular, its behaviour near the edges of the region over which the data have been collected is superior to that of the local mean approach. This is illustrated in the top right panel of Fig. 3.2, where the local linear and local mean estimators are contrasted, using a large value of smoothing parameter. The large bias of the local mean estimator at the edges of the covariate space is apparent. An outline of some of the technical issues will be given in Chapter 4.

The local linear estimator, with normal kernel functions, will be the standard technique used throughout this book. This approach has the further attractive property that it can be viewed as a relaxation of the usual linear regression model. As h becomes very large the weights attached to each observation by the kernel functions become increasingly close and the curve estimate approaches the fitted least squares regression line. It is appealing to have this standard model within the nonparametric formulation. The straight line displayed in Fig. 3.1 was created by setting the smoothing parameter to a very large value. An additional feature of the comparison with the local mean approach is that the latter converges to a straight line parallel to the x axis, with intercept $\bar{y}$, as the smoothing parameter becomes very large. This is a less appealing end point.

It would clearly be possible to increase the number of polynomial terms in (3.2) and so to derive other types of polynomial estimators. However, the local linear approach is sufficiently flexible for most purposes and it has a number of advantages which will be mentioned briefly in Chapter 4.

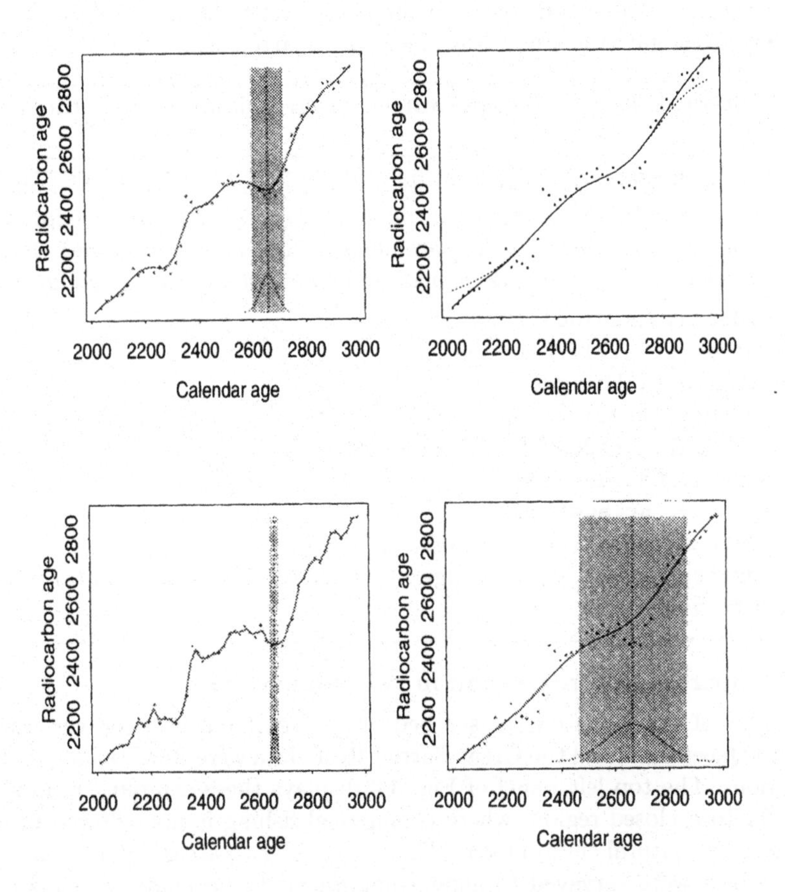

FIG. 3.2. Nonparametric regression curves for the radiocarbon data, using a variety of smoothing parameters. The top right panel contrasts the local linear (full line) and local mean (dashed line) estimators using a large value of smoothing parameter.

One particular feature of the radiocarbon data used in Figs 3.1 and 3.2 is that some information on the precision of each radiocarbon age is also available. It is straighforward to incorporate this information into the local linear estimator by attaching a further component to each kernel weight. The least squares problem then becomes

$$\min_{\alpha,\beta} \sum_{i=1}^{n} \{y_i - \alpha - \beta(x_i - x)\}^2 \, w(x_i - x\,; h) \, \frac{1}{p_i^2},$$

where p_i denotes the precision associated with observation i. Readers are invited to explore the effect of this on the estimator for the radiocarbon data in an exercise at the end of this chapter. Clark (1977) describes in some detail applications of nonparametric regression curves in the context of radiocarbon data. Scott *et al.* (1984) investigate the treatment of errors in particular.

S-Plus Illustration 3.1. Nonparametric regression on the radiocarbon data

The following S-Plus *code may be used to reconstruct Fig. 3.1. The effect of changing the smoothing parameter, as illustrated in Fig. 3.2, can be explored by altering the value assigned to the parameter* h. *The argument* panel=T *allows this to be done in an interactive manner and to be viewed as an animation.*

```
provide.data(radioc)
par(mfrow=c(1,2))
plot(Cal.age, Rc.age)
abline(0,1,lty=2)
ind <- (Cal.age>2000 & Cal.age<3000)
cal.age <- Cal.age[ind]
rc.age  <-  Rc.age[ind]
sm.regression(cal.age, rc.age, h = 30)
sm.regression(cal.age, rc.age, h = 1000, lty = 2, add=T)
par(mfrow=c(1,1))
```

3.3 Nonparametric regression in two dimensions

In a survey of the fauna on the sea bed in an area lying between the coast of northern Queensland and the Great Barrier Reef, data were collected at a number of locations. The top left panel of Fig. 3.3 displays these sampling points. The dots refer to a closed region, where commercial fishing is not allowed in order to protect the natural environment. The circles refer to areas outside the closed region, which were surveyed to allow comparisons to be made. In view of the large numbers and types of species captured in the survey the response variable is expressed as a score, on a log weight scale, which combines information across species. Two such scores are available. The details of the survey, and a full analysis of the data, are provided by Poiner *et al.* (1997).

The two right hand panels of Fig. 3.3 show nonparametric estimates of the relationship between the catch score within the closed zone and the spatial coordinates, latitude and longitude, for the 1993 survey. There is little evidence of change with latitude. (Note that points of low latitude correspond to a corner of the closed region which deviates from the overall rectangular shape of the survey area.) There is, however, a marked change in the catch score with longitude. As the coastline here runs in a roughly north–south direction it is worthwhile exploring whether this effect is simply one of depth. This will be investigated in Sections 4.6 and 4.7.

In order to explore the data further it would be useful to describe the pattern of catch scores as a function of both latitude and longitude. The local linear approach can easily be extended to nonparametric regression in two dimensions. If the observed data are denoted by $\{x_{1i}, x_{2i}, y_i; i = 1, \ldots, n\}$, then for estimation at the point (x_1, x_2) the weighted least squares formulation, extending (3.2), is

$$\min_{\alpha,\beta,\gamma} \sum_{i=1}^{n} \{y_i - \alpha - \beta(x_{1i} - x_1) - \gamma(x_{2i} - x_2)\}^2 \, w(x_{1i} - x_1; h_1) \, w(x_{2i} - x_2; h_2).$$

Weights defined by a more general two-dimensional kernel function $w(x_{1i} - x_1, x_{2i} - x_2; h_1, h_2)$ could be used. However, the simpler construction as a product of two separate weight functions for each covariate is generally sufficient. The resulting estimator, corresponding again to the quantity $\hat{\alpha}$, can be compactly defined in matrix notation such as that adopted by Ruppert and Wand (1994). If X denotes an $n \times 3$ design matrix whose ith row consists of the elements $\{1 \ (x_{1i} - x_1) \ (x_{2i} - x_2)\}$, and W denotes a matrix of 0s with the weights $w(x_{1i} - x_1; h_1)\, w(x_{2i} - x_2; h_2)$ for each observation down the diagonal, then the local linear estimator can be written as the first element of the least squares solution $(X^\top W X)^{-1} X^\top W y$, where y denotes the vector of responses for each observation.

A two-dimensional nonparametric regression estimator is displayed in the bottom left panel of Fig. 3.3. This shows that the dominant effect on catch score is indeed one of longitude, with a steep shelf appearing around 143.4°E. There is little change with latitude at any value of longitude. In fact, the observed pattern can be associated with distance offshore and the corresponding nature of the sea bed.

S-Plus Illustration 3.2. Nonparametric regressions with the Great Barrier Reef data

---

The following S-Plus *code may be used to reconstruct Fig. 3.3.*

```
provide.data(trawl)
par(mfrow = c(2,2))
plot(Longitude, Latitude, type = "n")
points(Longitude[Zone == 1], Latitude[Zone == 1])
text(longitude[Zone == 0], Latitude[Zone == 0], "o")
Zone93    <- (Year == 1 & Zone == 1)
Position  <- cbind(Longitude - 143, Latitude)
sm.regression(Latitude[Zone93],  Score1[Zone93], h = 0.1)
sm.regression(Position[Zone93,], Score1[Zone93],
     h= c(0.1, 0.1), eye = c(8,-6,5), xlab="Longitude - 143")
sm.regression(Longitude[Zone93], Score1[Zone93], h = 0.1)
par(mfrow = c(1,1))
```

3.4 Local likelihood and smooth logistic regression

Figure 3.4 displays a small portion of data, presented by Hosmer and Lemeshow (1989), from a study of low birthweight in babies. Specifically, each letter 'S'

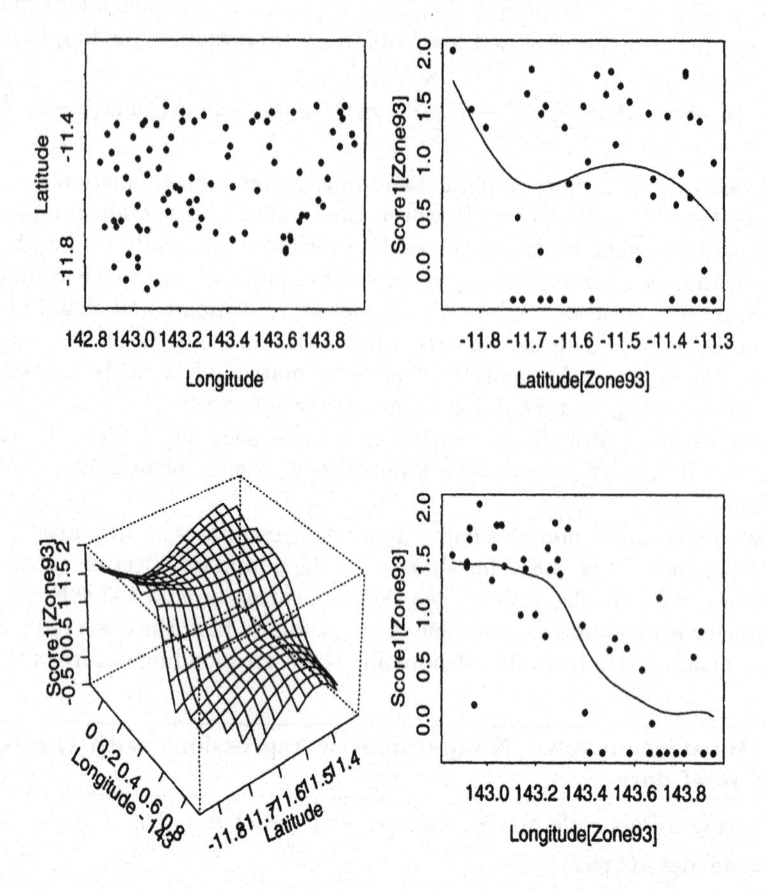

FIG. 3.3. Plots and nonparametric regression for the Great Barrier Reef data.

is placed at ordinate 1 or 0, depending on whether the birthweight is 'Low' or not, where 'Low' is defined as a weight below 2500 grams. The 0 and 1 values have been 'jittered' (i.e. their vertical position has been slightly perturbed) to limit overprinting and enhance the visual perception of the data. The abscissa of the 'S' gives the weight of the mother in pounds. The continuous line will be explained shortly.

It is natural to examine the relationship between the mother's weight and the probability of a low birthweight. A simple scatterplot of the data is particularly uninformative in this case, and some alternative form of plot must be introduced for graphical inspection of any underlying relationship. There is a variety of possibilities. One simple option is to use $\hat{m}(x)$, defined by (3.3), since the data consist of an explanatory variable x, namely the mother's weight, and a response variable y, the 0–1 low birthweight indicator. Since $\hat{m}(x)$ estimates

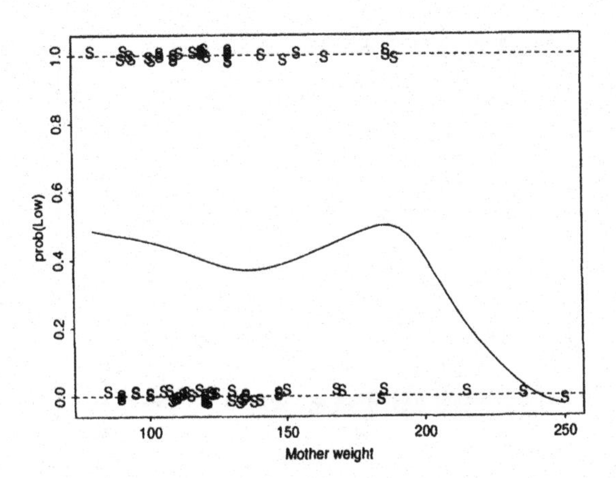

FIG. 3.4. A plot of data on the occurrence of low birthweight, with a smooth curve, obtained by local linear regression with $h = 20$, superimposed.

$\mathbb{E}\{y|x\}$, it therefore gives an estimate of $\mathbb{P}\{\text{Low}|x\}$ in this case. This produces the continuous line plotted in Figure 3.4, computed with smoothing parameter $h = 20$.

While the outcome of this method is numerically plausible, at least for the example displayed in Fig. 3.4, there are some obvious objections to it. The principal one is the fact that the binary nature of the response variable has been ignored. In particular,

- The motivation behind $\hat{m}(x)$ was provided by (3.1), which cannot hold in the present case, if ε is to be an error term in the traditional sense.
- The fitted curve is not guaranteed to lie in the interval $(0, 1)$. For example, the right tail of the curve plotted in Fig. 3.4 falls below 0, which is a highly undesirable feature for an estimated probability.
- The method does not naturally lend itself to satisfactory evaluation of standard errors, when these are required. Once again, this occurs because the binary nature of the response variable has not been taken into account.

To overcome these difficulties, a generalisation of the previous methodology is needed, to allow binary and other types of data to be handled satisfactorily. A natural extension of the weighting mechanism of (3.2) is to apply weights to the log-likelihood of the form

$$\ell_{[h,x]}(\alpha, \beta) = \sum_i \ell_i(\alpha, \beta)\, w(x_i - x; h), \tag{3.4}$$

where $\ell_i(\alpha, \beta)$ is the contribution to the usual log-likelihood from the ith observation; that is,

$$\ell_i(\alpha, \beta) = y_i \log\left(\frac{p_i}{1 - p_i}\right) + \log(1 - p_i).$$

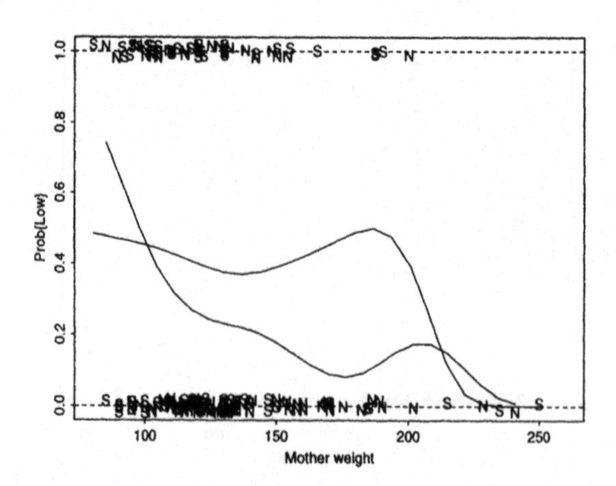

FIG. 3.5. A plot of the low birthweight data, and smooth curves obtained by local logistic regression with $h = 20$ for both the smoking (S) and non-smoking (N) groups of mothers.

Here p_i denotes the probability of a 1 at design point x_i, and as usual the logit link function

$$\text{logit}(p_i) = \log\left(\frac{p_i}{1-p_i}\right) = \alpha + \beta x_i \qquad (i = 1, \dots, n)$$

is assumed to relate p_i to a linear predictor in x_i.

Equation (3.4) defines a form of *local likelihood*, obtained by summing the contribution ℓ_i from each observation, weighted by the distance between the corresponding x_i and the point of estimation x. Maximisation of $\ell_{[h,x]}(\alpha, \beta)$ with respect to (α, β) provides local estimates $(\hat{\alpha}, \hat{\beta})$ and a fitted value

$$\hat{m}(x) = \frac{\exp(\hat{\alpha} + \hat{\beta}x)}{1 + \exp(\hat{\alpha} + \hat{\beta}x)}.$$

For simplicity, this notation does not explicitly indicate the dependence of the $(\hat{\alpha}, \hat{\beta})$ on x and h. By repeating this process for a sequence of x values, a set of pairs of the form $(x, \hat{m}(x))$ can be constructed and plotted.

Figure 3.5 was obtained by applying this procedure to the data of Fig. 3.4 and to a similar set, marked by the letter 'N'. The two groups of data differ in that the label S indicates that the mother is a smoker, and N indicates that she is not.

When interpreting Fig. 3.5, it is helpful to consider a traditional parametric logistic regression model of the form

$$\text{logit}\,(\mathbb{P}\{\text{Low}\}) = \alpha + \beta_1\,\text{Lwt} + \beta_2\,\text{Smoke}\,.$$

This corresponds to two fitted curves, one for each of the two groups indexed by the factor 'Smoke'. The estimate of β_2 is significant at the 5% level, although not markedly so, with an observed t ratio equal to 2.09.

A natural question is whether this significant difference between the two groups is present across the entire range of maternal weights or whether it applies only in some particular region, such as above or below a certain threshold. In fact, in a practical data analysis problem, a parametric fit may often be obtained first, and a graph such as Fig. 3.5 produced to examine where on the x axis any differences may occur. In Fig. 3.5, the curves are clearly far apart only in the central part of the abscissa. To assess whether the observed differences are significant, the associated standard errors would provide useful insight. In Chapter 6, a method which uses these to compare two curves will be discussed.

The concept of local likelihood lends itself easily to a variety of extensions.

- If two continuous covariates are present, x_1 and x_2 say, the local likelihood (3.4) is modified to

$$\ell_{[h_1,h_2,x_1,x_2]}(\alpha,\beta_1,\beta_2) = \sum_i \ell_i(\alpha,\beta_1,\beta_2)\, w(x_{1i}-x_1;h_1)\, w(x_{2i}-x_2;h_2)$$

  and the logistic assumption is extended to

$$\log\left(\frac{p_i}{1-p_i}\right) = \alpha + \beta_1 x_{1i} + \beta_2 x_{2i}.$$

- By suitably modifying the form of the ℓ_i components of (3.4), other types of data can be handled, for example binomial with index possibly greater than 1, Poisson or gamma.
- It is also possible to replace the logistic assumption by an alternative link function, but this is unlikely to produce any relevant difference in the final outcome.

As a second example, data reported by Hand *et al.* (1994, p. 65) are considered. These refer to the skeletal muscle of rats, whose fibres can be classified as 'Type I' or 'Type II', with Type I fibres further subdivided into three categories. The purpose of the study which generated the data was to 'set up a model which relates the number of Type II fibres to the counts of the three different types of Type I fibre'. However, for simplicity, the discussion below ignores the sub-categories within Type I.

A natural approach is to introduce a generalised linear model with a Poisson distribution for the Type II counts. Since the canonical link for the Poisson case is logarithmic, the model with linear predictor

$$\log \mathbb{E}\{\text{Type II}\} = \alpha + \beta \log(\text{Type I})$$

emerges directly. The plausibility of this relationship is also supported by a scatterplot of log(Type II) against log(Type I), which displays a clear linear pattern.

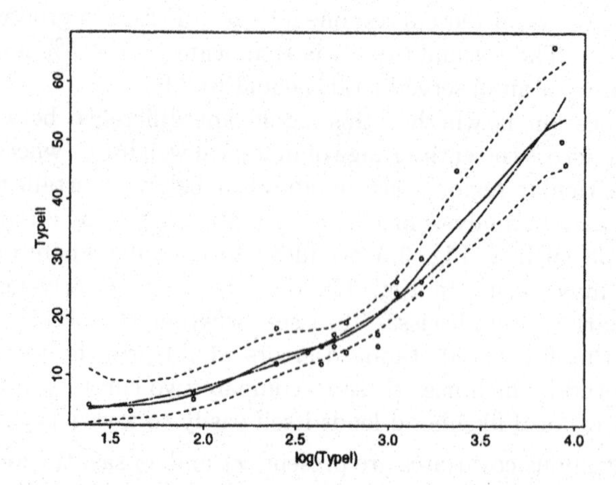

FIG. 3.6. A scatterplot of the rat muscle data, with nonparametric model (continuous line), an indication of its variability (dashed lines), and the log-linear parametric model (dot-dashed line) superimposed.

As an alternative graphical check, the link function between log (Type I) and $\mathbb{E}\{\text{Type II}\}$ can be examined with the aid of local likelihood, except that now the ℓ_i terms entering (3.4) are of Poisson form, rather than binomial as discussed earlier. This produces a nonparametric estimate of $\mathbb{E}\{\text{Type II}\}$ as a function of log (Type I), and the outcome can be compared with the parametric fit.

The result is displayed in Fig. 3.6, where the continuous curve represents the nonparametric estimate of $\mathbb{E}\{\text{Type II}\}$. The two dashed curves represent a measure of variability for the nonparametric estimate. The construction of these will be discussed in Section 4.4. The dot-dashed line shows the log-linear parametric model, which lies very close to the nonparametric one, in the light of the displayed measure of variability. This strengthens the support for the proposed parametric model.

In this example, the agreement between the parametric and the nonparametric estimates is very good across the whole span of the covariate axis, and further checks do not appear necessary. If the two curves differed appreciably at some points in the axis the interpretation of the plot would be less obvious, taking into account the local rather than global nature of the dashed bands. In this situation, the methods to be discussed in Chapter 5 provide a more formal tool for checking the fitted parametric model in a global manner.

S-Plus Illustration 3.3. Birthweight data – 1

Figure 3.4 was generated by the following S-Plus *code, which makes use of standard nonparametric regression for continuous data.*

```
provide.data(birth)
Low1<-Low[Smoke=="S"]; Lwt1<-Lwt[Smoke=="S"]
Lj <- jitter(Low1)
plot(Lwt1,Lj,type="n",xlab="Mother weight",ylab="prob(Low)")
text(Lwt1,Lj,"S",col=3)
abline(0,0, lty=3)
abline(1,0, lty=3)
sm.regression(Lwt1,Low1,h=20,add=T)
```

S-Plus Illustration 3.4. Birthweight data – 2

Figure 3.5 was generated by the following S-Plus *code, which makes use of local likelihood fitting for binary data.*

```
provide.data(birth)
Low0 <- Low[Smoke=="N"]; Lwt0 <- Lwt[Smoke=="N"]
Low1 <- Low[Smoke=="S"]; Lwt1 <- Lwt[Smoke=="S"]
sm.logit(Lwt0, Low0, h=20, pch="N", col=2,
      xlab="Mother weight", ylab="Prob{Low}", xlim=c(80,260))
sm.logit(Lwt1, Low1, h=20, pch="S", col=3, add=T)
```

S-Plus Illustration 3.5. Rat muscle data

Figure 3.6 was generated by the following S-Plus *code.*

```
provide.data(muscle)
TypeI <- TypeI.P+TypeI.R+TypeI.B
sm.poisson(log(TypeI), TypeII, 0.25, display="se")
pm <- glm(TypeII ~ log(TypeI), family=poisson)
line(log(TypeI), fitted(pm), col=4, lty=6)
```

3.5 Estimating quantiles and smoothing survival data

A nonparametric regression curve provides an estimate of $\mathbb{E}\{Y|x\}$. It can also be useful to construct estimates of conditional quantiles, rather than the conditional mean. These can be used to indicate 'normal ranges' in a nonparametric manner, or to highlight skewness or changes in variance as a function of x. If d_i denotes the deviations between the data and the local model, such as $y_i - \alpha - \beta(x_i - x)$, then one approach can be viewed as a modification of the usual forms of nonparametric regression by replacing the d_i^2 term in expressions such as (3.2) with the function $|d_i| + (2\alpha - 1)d_i$, to estimate the α-level quantile. This is described by Green and Silverman (1994) and Fan and Gijbels (1996, Section 5.5.2). In this section, an alternative approach, which allows a more direct method of evaluation, and which extends to the case of censored data, is described.

A famous set of data was collected in connection with a study of heart transplant patients where the survival times of patients were recorded along with a

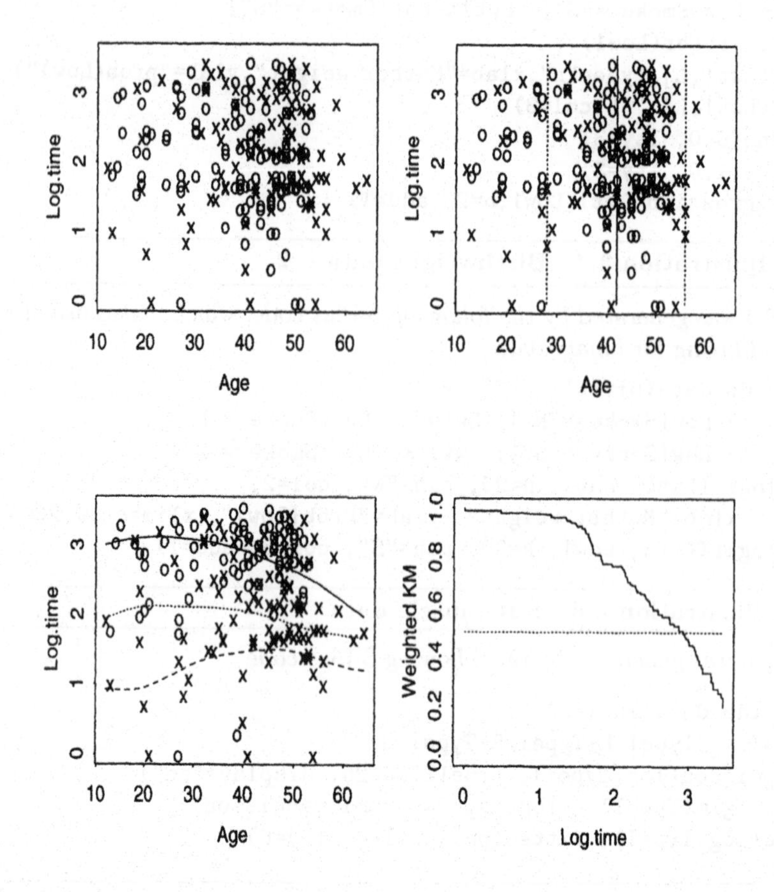

FIG. 3.7. The construction of smooth percentile curves for the Stanford heart transplant data. The top left panel contains a scatterplot of the data with censored observations identified as 'o'. The top right panel shows the approximate span of a kernel function centred at age 42. The bottom right panel shows a locally weighted Kaplan–Meier survival plot. The bottom left panel shows smooth estimates of percentiles curves for $p = 0.5, 0.25, 0.1$.

variety of covariates. Figure 3.7 shows a scatterplot of these Stanford heart transplant data, displaying survival times against the age of the patients. Since the survival times have a markedly skewed distribution a log scale has been used. Times corresponding to patient deaths are marked by '×' while times corresponding to censored data, where the patient remained alive at the last time of monitoring, are marked by 'o'.

It can be difficult to interpret scatterplots of survival data because of the different information carried by the death times and the censored observations.

Standard models for this type of data express relationships on the hazard function scale. It would, however, be useful to be able to view evidence of a relationship between a covariate and response in an exploratory manner on a scatterplot with the original survival time scale, which is more easily interpreted. This might in addition be useful in checking the assumptions, such as proportionality of hazard functions, which many standard models adopt.

It is common to use medians to summarise groups of survival data, as the median can still be computed easily in the presence of censored data. The idea of a running median is therefore a useful one to indicate the nature of a relationship between a covariate and survival times. The top right panel of Fig. 3.7 displays a 'window' centred at age 42. The data which fall within this window can be extracted and a Kaplan–Meier survival curve constructed, from which the local median can be computed. Repeating this operation at a variety of covariate values would produce an indication of how the median changes as a function of age.

When it is reasonable to suppose that the underlying median curve is smooth, a kernel approach to this problem can be adopted. Beran (1981) was one of the first to propose this and its graphical application has since been developed by Doksum and Yandell (1982) and Gentleman and Crowley (1991). At the covariate value x, weights $w(x_i - x; h)$ are computed to allow the construction of a weighted Kaplan–Meier survival curve given by

$$\hat{S}(y) = \prod_{i: y_i < y} \left[1 - \frac{\delta_i \, w(x_i - x; h)}{\sum_{j \in R_i} w(x_j - x; h)} \right]^{\delta_i},$$

where δ_i is an indicator of death (1) or censoring (0) and R_i denotes the set of individuals at risk at time y_i. This curve is a step function, as shown in the bottom right panel of Fig. 3.7. The median can be found by a simple search. However, in order to produce a smooth curve, Wright and Bowman (1997) propose smoothing the step function by applying a nonparametric regression procedure to the data representing the corners of the steps. Since it is the time value at which the weighted survival curve takes the value 0.5 which is of interest, the vertical axis (proportion) is used as the covariate and the horizontal axis (survival time) as the response in the nonparametric regression. This allows an estimate of the median to be constructed in a simple and direct manner. The 'vertical' smoothing parameter required at this stage is relatively unimportant and can be set to a small value on the proportion scale, such as 0.05. The result of this operation is that the curve estimate is genuinely smooth, as shown in the bottom left hand panel of Fig. 3.7. In addition to the median, curves have also been computed for the 25th and 10th percentiles.

An interesting feature of the median curve produced by the Stanford data is that it suggests a decrease in survival time for higher values of age. This is not immediately obvious from the original scatterplot of the data where the overall pattern of points does not show an obvious trend with age, and it is difficult to identify visually the effect of the pattern of censored observations.

An alternative approach to the construction of a nonparametric curve for survival data is to fit a standard proportional hazards model in a local manner, from which a local median can be derived. Nonparametric models for the hazard function are described by Hastie and Tibshirani (1990, Section 8.3). For exploratory purposes, the approach described above has the advantage of requiring relatively simple calculations. On the other hand, modelling on the hazard scale has the benefit of allowing more general models with several covariates, incorporating both parametric and nonparametric effects, to be formulated.

Fan and Gijbels (1994) discuss nonparametric estimation of the mean function in the presence of censored data on the response variable.

S-Plus Illustration 3.6. Percentile curves for survival data

The following S-Plus *code may be used to reconstruct the bottom left panel of Fig. 3.7.*

```
provide.data(stanford)
sm.survival(Age, Log.time, Status, h = 7)
sm.survival(Age, Log.time, Status, h = 7, p = 0.25,
        add = T, lty = 2)
sm.survival(Age, Log.time, Status, h = 7, p = 0.10,
        add = T, lty = 3)
```

3.6 Variable bandwidths

The estimators described in the previous sections of this chapter have used the same smoothing parameter h in the weights attached to each observation (x_i, y_i). In some situations there can be advantages in using different smoothing parameters at different covariate values. For example, Fig. 3.8 displays data on the number of ovarian follicles, on a log scale, counted from sectioned ovaries of women of various ages. These data were reported by Block (1952; 1953), Richardson *et al.* (1987) and A. Gougeon, and were analysed by Faddy and Gosden (1996) in a study of the relationship between age and log count. Faddy and Jones (1997) also analyse these data. Most of the subjects are more than 30 years of age, with relatively few observations on children and young women. In addition, there is a long, gentle decline in the log counts over the earlier years, followed by a more rapid fall. In view of this, it would be appealing to use a large smoothing parameter at smaller ages, where the data are sparser and the curve appears to be linear, and to use a smaller smoothing parameter at larger ages, where the data are denser and the underlying trend appears to exhibit more rapid change.

A simple way to implement this is to employ a ***variable bandwidth*** which reflects the density of the design points through a nearest neighbour distance. Following the strategy for density estimation described in Section 1.7, the bandwidth used in the kernel function for estimation at the point x could be defined

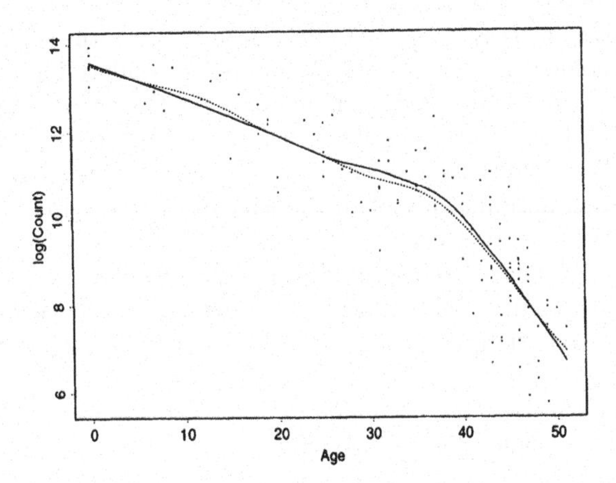

FIG. 3.8. A local linear estimator (dotted line) and `loess` estimator (full line) with the follicle data.

by $h_i = hd_k(x)/\bar{d}$, where $d_k(x)$ denotes the distance to the kth nearest neighbour of the covariate value x_i and $\bar{d}$ denotes the geometric mean of the $d_k(x)$. In this way, the overall bandwidth h is scaled to increase the degree of smoothing applied in regions where data are sparse, and to decrease the degree of smoothing where data are dense.

One of the earliest, and still very popular, approaches to nonparametric regression uses nearest neighbour distances in a particularly simple and appealing way. This was described by Cleveland (1979) and is referred to as the `lowess` estimator, or `loess` after its more general S-Plus implementation. The key component is that in the local linear estimator defined through the least squares criterion (3.2) the kernel function for estimation at the point x is $w(x_i - x; d_k(x))$. This achieves an appropriately variable pattern of smoothing without the need for an overall smoothing parameter h. To provide weights which are easily evaluated but adequately smooth, Cleveland (1979) used the tricube kernel function defined by $w(z;h) = (1 - (|z|/h)^3)^3$, for $z \in [-h, h]$. A further appealing feature of this formulation is that the degree of smoothing applied to the data can be expressed in the parameter k/n, which describes the proportion of the sample which contributes positive weight to each local linear regression. This is referred to as the ***span*** of the estimator and it has the attraction of a clear and simple interpretation.

Figure 3.8 displays a `loess` estimator, using the default span of 0.5, and a standard local linear estimator with constant bandwidth $h = 4$, on a plot of the follicle data. The differences between the estimators are not very substantial on these data. However, the larger bandwidths in the `loess` estimator for younger women produce a straighter curve over the period of gentle decline, and the smaller bandwidths used with the data on older women allow the sharp change in the rate of decline to be tracked a little more effectively.

This approach is successful in the follicle data because the higher density of covariate values coincides with the region where the underlying curve changes most rapidly. For data where rapid changes in the regression curve occur in regions where data have been sampled sparsely it would be less appropriate to employ variable bandwidths which were based on nearest neighbour distances, and which thus took account only of the density of the design points.

The `lowess` estimator proposed by Cleveland (1979) also incorporates robustness in the fitting procedure, to prevent unusual observations from exerting large influence on the fitted curve. This can be implemented by the use of appropriate weights in an iterative smoothing procedure, in a manner which parallels robust linear regression.

For further description and discussion of the `loess` estimator, see Cleveland and Devlin (1988) and Cleveland *et al.* (1992).

S-Plus Illustration 3.7. Variable bandwidths in nonparametric regression

---

The following S-Plus *code may be used to reconstruct Fig. 3.8.*

```
provide.data(follicle)
sm.regression(Age, log(Count), h = 4, lty = 2)
model <- loess(log(Count) ~ Age)
lines(Age, model$fitted.values, col = 6)
```

3.7 Alternative forms of nonparametric regression

3.7.1 *Gasser–Müller estimate*

An alternative form of kernel estimator, where the amount of weight attached to each observation locally is determined by the spacings between the design points, was proposed by Gasser and Müller (1979). This estimator is defined by

$$\hat{m}_{GM}(x) = \sum_{i=1}^{n} y_{[i]} \int_{t_{i-1}}^{t_i} \frac{1}{h} w\left(\frac{x-y}{h}\right) dt,$$

where $t_0 = -\infty$, $t_i = \frac{1}{2}(x_{[i]} + x_{[i+1]})$, $t_n = \infty$, $x_{[i]}$ denotes the ith largest value of the observed covariate values and $y_{[i]}$ denotes the corresponding response value. This approach has some attractive properties. However, it has been shown that the variance of the estimator can be rather higher than that of other methods under some circumstances.

3.7.2 *Smoothing splines*

Splines represent a well established mathematical tool, familiar in particular to numerical analysts, who use splines to fit a function through a sequence of points on the plane. The term 'spline' is borrowed from a mechanical device which was used to draw cross-sections of ships' hulls. The mechanical spline was a flexible strip of wood which was forced to pass through certain fixed points, and otherwise allowed to find its natural position.

In mathematics, 'spline' denotes a function $s(x)$ which is essentially a piecewise polynomial over an interval (a, b), such that a certain number of its derivatives are continuous for all points of (a, b). More precisely, $s(x)$ must satisfy the following properties. For some given positive integer r and a sequence of points $t_1, \ldots, t_k$, called *knots*, such that $a < t_1 < \ldots < t_k < b$, it is required that:

- for any subinterval (t_j, t_{j+1}), $s(x)$ is a polynomial of order r;
- $s(x)$ has $r - 2$ continuous derivatives;
- the $(r-1)$th derivative of $s(x)$ is a step function with jumps at $t_1, \ldots, t_k$.

Often r is chosen to be 3, and the term *cubic spline* is then used for the associated curve.

In statistics, splines are used not for data interpolation, as in numerical analysis, but for data smoothing. Specifically, the quantity

$$D = \sum_{i=1}^{n} (y_i - m(x_i))^2 + \lambda \int_a^b m''(x)^2 dx,$$

where λ is some positive constant, is used as an objective function in a minimisation problem with respect to the unknown regression function $m(x)$ on the basis of the data (x_i, y_i). The connection between D and splines is that the mathematical function which minimises D is a cubic spline, whose specific expression depends on the data and on the choice of λ.

If $\lambda = 0$, minimisation of D corresponds to interpolation of the data, which is not a useful method of fitting the model (3.1), since the residuals are all set equal to 0. To avoid this, the second term in the expression of D is then inserted as a *roughness penalty*. The choice of λ determines the relative weight attributed to the two terms, the residual sum of squares and the roughness penalty $\int m''(x)^2 dx$. Increasing λ penalises fluctuations, and so produces a smoother curve. The term *penalised least squares* is then used in conjunction with D. If $\lambda \to \infty$, the second derivative is effectively constrained to be 0, and the outcome is then the least squares line. Hence, λ plays a similar role to the smoothing parameter h used in earlier sections.

The philosophy underlying this approach is an elegant and powerful one and it can be modified to apply to many other situations. Extensive discussions are given by Eubank (1988), Wahba (1990) and Green and Silverman (1994).

3.7.3 *Orthogonal series and wavelets*

The orthogonal series or projection approach, mentioned in Section 1.7 in connection with density estimation, extends to the present setting by taking the function f expanded in (1.3) to be the regression function $m(x)$. An account of the use of this approach employing the trigonometric functions or Legendre polynomials as a basis is given by Eubank (1988, Chapter 3).

However, much attention is currently being directed towards a different form of estimator which, although still falling under the umbrella of 'orthogonal series estimators', exhibits rather different behaviour, and can therefore be considered as a distinct approach. Here, a doubly indexed family of functions $\{\psi_{jk}(x)\}$ leads to the representation of a function f in the form

$$f(x) = \sum_j \sum_k c_{jk}\psi_{jk}(x),$$

where $c_{jk} = \int_{\mathbb{R}} f(x)\, \psi_{jk}(x)dx$ are the coefficients of the projection.

The novel feature is the use of particular functions, called *wavelets*, for the basis. These take the form

$$\psi_{jk}(x) = 2^{j/2}\psi(2^j x - k)$$

for some given function $\psi(x)$, called the *mother wavelet*, with additional requirements to ensure a basis is created, as described in Section 1.7. The whole set of functions is therefore generated from the mother wavelet, by applying a scale and a shift transform.

Clearly, the coefficients c_{jk} must be estimated from the data and only a finite set of them is retained in the estimator, in a similar manner to (1.4). The number of j indices considered for each given k does not need to be constant. This fact, coupled with the special nature of the wavelet bases, where the index k is associated with location and the index j with frequency, allows variation in the degree of smoothness along the x axis. The term *multiresolution analysis* is used in this context. This important and powerful feature is one of the principal reasons for the current research interest in this topic. Methodological and practical issues are still under development.

An accessible account of the wavelet approach is provided by Fan and Gijbels (1996, Section 2.5).

3.7.4 *Discussion*

The relative merits of the different forms of nonparametric regression estimators have been well discussed by other authors. It is not the aim of this text to advocate one method over another but to illustrate the applications of nonparametric regression and to describe some of the tools which may be derived from it. Throughout the remainder of this book attention will be focused substantially on the local linear approach to nonparametric regression. An advantage of the local linear approach is that it can be extended without conceptual difficulty to two dimensions, and that the use of kernel functions to define a local weighting provides a natural link with density estimation. However, many of the methods to be described in the remainder of this book can also be applied with other smoothing techniques.

3.8 Further reading

Useful overviews of different approaches to the construction of nonparametric regression estimators are given by Hastie and Tibshirani (1990, Chapter 2) and Fan and Gijbels (1996, Chapter 2). In addition, Chu and Marron (1991) and Jones *et al.* (1994) discuss various forms of estimators based on kernel functions. For the local linear approach in particular, recent discussion is provided by Hastie and Loader (1993) and Cleveland and Loader (1995).

For local likelihood, one formulation is described by Tibshirani and Hastie (1987). The discussion of Section 3.4 follows the development of Fan *et al.* (1995) and Wand and Jones (1995, pp. 164–167) more closely. This has the advantage of avoiding the need to handle in a special way those evaluation points x which do not belong to the observed set of covariate values.

The strategy of binning the data can be used to evaluate nonparametric regressions in a very efficient manner. Wand and Jones (1995, Appendix D), Fan and Gijbels (1996, Section 3.6) and Fan and Marron (1994) describe this in some detail.

Exercises

3.1 *The precision of radiocarbon ages.* In addition to the data used to construct Figs 3.1 and 3.2, the `radioc` file contains information on the `Precision` of the radiocarbon age. Amend the script given in Section 3.2 to define the precisions for the range of calendar ages between 2000 and 3000 years as

```
p <- Precision[ind]
```

and then use this to define additional weights in a local linear estimator by adding the parameter `weights = 1/p^2` to the `sm.regression` function. Do the weights make any difference in this case?

3.2 *Trawler data from the Great Barrier Reef.* A script to produce a two-dimensional nonparametric regression estimate is given in Section 3.3. Amend this script to produce a similar plot from the data for 1992 and so identify whether the pattern is similar in the two different years. Repeat this process for the regions outside the closed zone. Are there any obvious differences between the years or between the regions?

3.3 *Kernel function shape.* The S-Plus function `ksmooth` allows local mean regression estimates to be constructed from different shapes of kernel functions. For example, the commands

```
x <- runif(50)
y <- rnorm(50)
plot(x, y, type="n")
lines(ksmooth(y, x, bandwidth=1))
```

will plot a density estimate using a 'box' kernel by default. Compare the effects of using other kernel shapes, with the same value of bandwidth, by adding to `ksmooth` the argument `kernel` with the values `"triangle"`, `"parzen"` and `"normal"`.

3.4 *Derivation of the local linear estimator.* Verify that the expression given in (3.3) for the local linear estimator is in fact the solution to the weighted least squares problem defined by (3.2).

3.5 *Animated smoothing of survival curves.* The file `stananim.S` contains S-Plus code to create an animated plot of a smooth survival curve by repeated erasure and replotting, using a new value of smoothing parameter. Use this code to explore the effect of changing the smoothing parameter when constructing a median curve for the Stanford data.
In Section 3.5 it was suggested that the value of the 'vertical' smoothing parameter is relatively unimportant. Modify the code contained in `stananim.S` to allow animation over this smoothing parameter, which is controlled in the function through the argument `hv`. Do you agree that this parameter has relatively little effect?

3.6 *Animation of `loess`.* Exercise 3.5 refers to the creation of an animated plot for a nonparametric regression curve for survival data. Repeat this for the function `loess` in the usual case of uncensored data, to explore how the shape of the curve changes with the `span` parameter.

3.7 *Circular regression.* When the covariate x in a regression problem is periodic, referring for example to days or weeks throughout the year, the techniques for creating circularity, described in the context of density estimation in Section 1.5, can be applied. Write an S-Plus function which will implement this for nonparametric regression by replicating the observed data over regions of the x axis corresponding to the periods immediately below and above the range over which the nonparametric estimate will be evaluated.

3.8 *Worm data.* In a study of the parasitic infection *Ascaris lumbricoides*, Weidong *et al.* (1996) report on a survey of a rural village in China. The file `worm` contains an indicator of the presence or absence of the parasite in faecal samples from males and females of different ages. Use `sm.logit` to plot nonparametric estimates of the probability of occurrence of the parasite as a function of age, separately for males and females. Comment on the shapes of these curves.

3.9 *Follicle data.* Construct a variability band for the regression curve and identify the features of the data which cause the changes in the width of the curve. Construct a density estimate of age to show how the density of design points changes.

4

INFERENCE WITH NONPARAMETRIC REGRESSION

4.1 Introduction

Some basic ideas of nonparametric regression were introduced in Chapter 3 and illustrated with a variety of data of different types. The emphasis has so far been on the graphical representation of data. However, a number of issues have arisen. A principal question is how the features of a nonparametric regression curve can be assessed, to discover whether these indicate structural effects or whether they can be attributed simply to random variation. In addition, some attention needs to be devoted to methods of selecting suitable smoothing parameters. In order to prepare the way for a discussion of these issues some very simple properties of nonparametric regression estimators are derived. Attention is restricted mostly to the case where the response y and a covariate x both lie on continuous scales. Many theoretical results exist in this area. However, for the purposes of the tools to be described in later chapters it is sufficient at this stage to focus simply on means and variances.

Techniques of inference for a variety of different problems form the main focus of most of the remaining chapters of this book. The aim of the present chapter is therefore to set the scene with some introductory ideas.

4.2 Simple properties

From a review of the forms of nonparametric regression estimators with continuous data described in Section 3.2 and 3.3 it becomes apparent that these are linear in the observed responses y_i. This general structure, which can be expressed as $\hat{m}(x) = \sum v_i y_i$ for some known weights v_i, is very convenient in a number of ways. It is particularly helpful when exploring the simple properties of the estimators, such as means and variances, since these can be computed very easily as

$$\mathbb{E}\{\hat{m}(x)\} = \sum_{i=1}^{n} v_i m(x_i), \tag{4.1}$$

$$\text{var}\{\hat{m}(x)\} = \left(\sum_{i=1}^{n} v_i^2\right)\sigma^2. \tag{4.2}$$

Note in particular that it is possible to compute the mean simply by smoothing the points $(x_i, m(x_i))$ rather than the points (x_i, y_i).

Figure 4.1 displays the curve $\sin(2\pi x)$ over the range $(0, 1)$, with 50 simulated observations superimposed. The standard deviation of the error distribution is

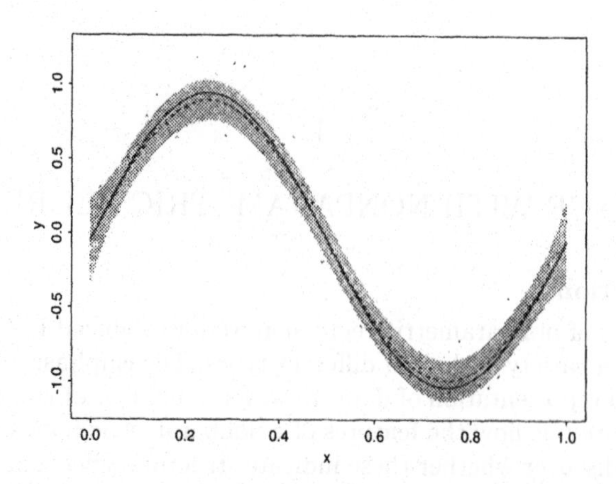

FIG. 4.1. The mean (dotted line), and a shaded area representing the mean plus or minus two standard deviations, for a local linear nonparametric regression estimator based on data from a sine curve (full line).

0.2. The dashed line displays the mean value of the local linear estimator and the shaded band indicates the region corresponding to two standard deviations of the estimate about its mean. There is an apparent bias at peaks and troughs of the function.

Graphical exploration of this type is helpful in appreciating the nature and size of the bias and variance effects. It is also instructive to derive a mathematical description. General expressions for these can be found in Ruppert and Wand (1994) for the local linear case. For a one-dimensional covariate they are

$$\mathbb{E}\{\hat{m}(x)\} \approx m(x) + \frac{h^2}{2}\sigma_w^2\, m''(x), \tag{4.3}$$

$$\text{var}\{\hat{m}(x)\} \approx \frac{\sigma^2}{nh}\frac{\alpha(w)}{f(x)}, \tag{4.4}$$

where σ_w^2 denotes $\int z^2 w(z)dz$ and $\alpha(w)$ denotes $\int w(z)^2 dz$. An informal derivation of these results is provided at the end of this section. In order to consider asymptotic expressions of this type it is necessary to make some assumption about how the pattern of design points x_i changes as the sample size increases. This is assumed to be expressed in the density function $f(x)$. When the x values are chosen randomly from some distribution then f represents the corresponding probability density function. More generally, f can be taken to characterise the local density of design points as the sample size increases.

These asymptotic expressions therefore show that bias increases with larger smoothing parameters and with curves showing higher degrees of curvature m''. They also show that variance increases with smaller smoothing parameters and at values of x where neighbouring design points are scarce, since the factor

$nhf(x)$ can be viewed as controlling the local sample size. An appropriate balance between bias and variance as a function of the smoothing parameter is therefore required.

It is a very attractive property of the local linear approach to smoothing that the bias component does not depend on the pattern of the design points, at least asymptotically. Fan (1992; 1993) referred to this property as ***design adaptive*** and demonstrated other attractive theoretical properties of the estimator, including good behaviour near the extremes of the design points. These properties are not shared by the local constant approach and so the local linear approach has much to offer.

Some difficulties can arise, however. If a small smoothing parameter is used, and the design points contain a region where few observations have been taken, then it is possible that the estimator is undefined in some places. This general issue is discussed by Cheng *et al.* (1997). The variable bandwidth approach described in Chapter 3 also provides a simple way of addressing the problem of sparsity of the design points. The use of nearest neighbour distances in particular ensures that each kernel function spans a similar number of observations. It was shown in Chapter 2 that a nearest neighbour distance is proportional to a simple estimate $\tilde{f}$ of the underlying density function. A variable bandwidth of this type is therefore equivalent to $h/\tilde{f}(x_i)$, and so, from (4.4), this approach attempts to cancel the effect of the factor $f(x)$ in the denominator of the expression for the variance of the estimator. An unfortunate side-effect is that the bias becomes dependent on the design density $f(x)$; see Hastie and Tibshirani (1990, Section 3.10).

These bias and variance expressions provide further guidance on the role of variable bandwidth smoothing procedures, in addition to the effect of the density of design points. For example, local bandwidths should be small when curvature is large. A further issue arises if the error standard deviation changes as a function of x, with a consequent effect on the appropriate balance between bias and variance. Small bandwidths are appropriate when the standard deviation is small. Sometimes these features will occur together, as in the follicle data discussed in Chapter 3. However, when the adjustments required by these features conflict, a more careful choice of local bandwidths than simple nearest neighbours may be required. Cleveland and Loader (1995) and discussants provide a variety of views on this issue.

Throughout this book, the local linear approach to smoothing, with constant bandwidth, will be the standard method. An important factor in this choice is the design adaptive property described above, where bias is independent of the design density $f(x)$. This property is particularly useful when comparing models, which is an important feature of subsequent chapters.

It is possible to extend the local linear approach defined by (3.2) to other local polynomial functions, such as a local quadratic. Fan and Gijbels (1996, Section 3.3) discuss this issue and show that there are theoretical reasons for preferring odd powers of local polynomials. This strengthens the case for the use of the simple local linear estimator, although there are some situations where a local polynomial of higher degree may still be useful.

The mean and variance approximations (4.3) and (4.4) extend to the case of several predictor variables. The general expressions are given by Ruppert and Wand(1994). For later work here it is sufficient to consider the case of two covariates x_1 and x_2, whose estimator is defined in Section 3.3, using simple product kernel weights $w(x_{1i}-x_1;h_1)\,w(x_{2i}-x_2;h_2)$. In this case the approximate mean and variance are

$$\mathbb{E}\{\hat{m}(x_1,x_2)\} \approx \frac{1}{2}\sigma_w^2 \left\{\sum_{j=1}^{2} h_j^2 \frac{\partial^2}{\partial x_j^2} m(x_1,x_2)\right\},$$

$$\mathrm{var}\{\hat{m}(x_1,x_2)\} \approx \frac{\sigma^2}{nh_1h_2} \frac{\alpha(w)^2}{f(x_1,x_2)}.$$

Fan *et al.* (1995) also analyse nonparametric estimates in the context of local generalised linear models and derive analogous expressions for bias and variance in that setting.

In order to carry out inference with nonparametric regression curves it is necessary to consider their distributional properties. When the response and covariate are both on continuous scales, the error variance σ^2 is an important parameter. Estimators for this quantity will be discussed in the next section. However, most of the important distributional properties of nonparametric regression estimators follow from their characterisation as linear functions of the response variables y_i. For example, if it is appropriate to assume that the response variables are normally distributed then this linear structure ensures that the estimator is also normally distributed.

The linearity of the estimator can often be most conveniently expressed in vector-matrix notation as

$$\hat{m} = Sy, \tag{4.5}$$

where $\hat{m}$ denotes the vector of estimates at a set of evaluation points of interest, S denotes a *smoothing matrix* whose rows consist of the weights appropriate to estimation at each evaluation point, and y denotes the observed responses in vector form. Note that in order to construct fitted values $\hat{m}(x_i)$ the evaluation points are the observed covariate values x_i. However, the notation of (4.5) is more general and allows evaluation over an arbitrary set of x values, for example in a regularly spaced grid, by suitable modification of the weights in the rows of S.

Mathematical aspects: The approximate mean and variance of the local linear estimator

---

In the local linear case, the estimator (3.3) can be written in the form $\hat{m}(x) = \sum a_i y_i / \sum a_i$, where $a_i = n^{-1} w(x_i - x; h)\{s_2 - (x_i - x)s_1\}$. The quantity s_1 is defined, and then approximated in an informal way, by

$$\begin{aligned} s_1 &= \frac{1}{n}\sum_j w(x_j - x; h)\,(x_j - x) \\ &\approx \int w(u - x; h)\,(u - x)\,f(u)\,du \\ &\approx \int w(z)\,hz\,\{f(x) + hz\,f'(x)\}dz \\ &= h^2 f'(x), \end{aligned}$$

where, for simplicity, it is assumed that the kernel function is scaled to have variance 1. These steps arise from the convergence of the average to an integral weighted by $f(u)$, a change of variable and a Taylor expansion. By a similar argument,

$$s_2 \approx h^2 f(x).$$

This means that the weights a_i can be approximated by

$$a_i \approx \frac{1}{n} w(x_i - x; h) h^2 \{f(x) - (x_i - x) f'(x)\}.$$

The mean of the estimator is $\mathbb{E}\{\hat{m}(x)\} = \sum a_i m(x_i)/\sum a_i$. Ignoring the constant factor h^2 which cancels in the ratio, the numerator of this expression can then be approximated as

$$\left\{f(x)^2 + \frac{h^2}{2} f(x) f''(x) - h^2 f'(x)^2\right\} m(x) + \frac{h^2}{2} f(x)^2 m''(x),$$

after another integral approximation, change of variable and Taylor series expansion. By a similar argument, the denominator of $\mathbb{E}\{\hat{m}(x)\}$ can be approximated by

$$f(x)^2 + \frac{h^2}{2} f(x) f''(x) - h^2 f'(x)^2,$$

from which the principal terms of the ratio can be identified as

$$m(x) + \frac{h^2}{2} m''(x).$$

In the case of the variance, where $\mathrm{var}\{\hat{m}(x)\} = \sigma^2(\sum a_i^2)/(\sum a_i)^2$, similar arguments show the principal terms of the numerator and denominator are respectively $\sigma^2 f(x)^3 \alpha(w)/nh$ and $f(x)^4$. The ratio of these expressions produces the variance approximation (4.4).

4.3 Estimation of σ^2

In order to complete the description of a nonparametric regression model it is necessary to provide an estimate of the error variance parameter, σ^2. In a standard linear model the residual sum of squares provides a basis for this and an analogue in a nonparametric setting is

$$\text{RSS} = \sum \{y_i - \hat{m}(x_i)\}^2.$$

Using the matrix notation introduced in the previous section, the fitted values $\hat{m}$ are given by $\hat{m} = Sy$, where the evaluation points are the design points x_i. The matrix S is analogous to the projection, or *hat*, matrix $H = X(X^\top X)^{-1} X^\top$ in a standard linear model of the form $y = X\theta + \varepsilon$. Just as the degrees of freedom for error in a linear model can be expressed as $\text{tr}(I - H)$, so an analogous definition of *approximate degrees of freedom* in a nonparametric setting is $\text{df}_{\text{error}} = \text{tr}(I - S)$, where tr denotes the trace operator. An alternative is provided by $\text{tr}(I - S^\top)(I - S)$ on the basis of a different analogy. Whichever definition is used, the quantity

$$\hat{\sigma}^2 = \text{RSS}/\text{df}_{\text{error}} \tag{4.6}$$

then becomes a natural estimate of σ^2.

The analogy with a linear model is complicated by the presence of bias in the estimator $\hat{m}$, which will have the effect of inflating the size of the residual sum of squares. A possible solution is to reduce this bias by employing a small smoothing parameter for the purposes of estimating σ^2. An alternative approach was taken by Rice (1984) who proposed an estimator based on differences of the response variable, in order to remove the principal effects of the underlying mean curve. Specifically, the estimator is

$$\hat{\sigma}^2 = \frac{1}{2(n-1)} \sum_{i=2}^{n} (y_i - y_{i-1})^2, \tag{4.7}$$

where, for simplicity of notation, it is assumed that the observations (x_i, y_i) have been ordered by x_i.

One disadvantage of this approach is that the differences $y_i - y_{i-1}$ are influenced by the fluctuation in the underlying regression function and so the estimate of σ^2 will be inflated. Gasser *et al.* (1986) proposed an alternative method which removes much of this difficulty by using a form of linear interpolation. Specifically, pseudo-residuals are defined as

$$\tilde{\varepsilon}_i = \frac{x_{i+1} - x_i}{x_{i+1} - x_{i-1}} y_{i-1} + \frac{x_i - x_{i-1}}{x_{i+1} - x_{i-1}} y_{i+1} - y_i,$$

which measures the difference between the y_i and the line which joins its two immediate neighbours. If $\tilde{\varepsilon}_i$ is written in the form $a_i y_{i-1} + b_i y_{i+1} - y_i$, then the estimator of error variance is

$$\hat{\sigma}^2 = \frac{1}{n-2} \sum_{i=1}^{n-1} c_i^2 \tilde{\varepsilon}_i^2,$$

where $c_i^2 = (a_i^2 + b_i^2 + 1)^{-1}$. This approach substantially removes the influence of the underlying regression function on the estimator, and it will be used in later sections of this book. Further discussion is provided by Hall and Marron (1990), Hall *et al.* (1990) and Seifert *et al.* (1993).

When there are two covariates, approaches based on differencing become more awkward and estimators based on a residual sum of squares and approximate degrees of freedom are easier to construct.

4.4 Confidence intervals and variability bands

In all forms of statistical modelling, confidence intervals are extremely helpful in providing an indication of the uncertainty associated with estimates of quantities of interest. Most of the necessary ingredients for a confidence interval for the value $m(x)$ of a nonparametric regression curve at a specific point x are readily available. It was pointed out in Section 4.2 that when it is appropriate to assume that the error variance has a normal distribution then the nonparametric estimator $\hat{m}(x)$ will also be normally distributed. Even where the error distribution cannot be assumed to be normal, mild assumptions allow a normal approximation to be employed for $\hat{m}(x)$ by appealing to a form of the central limit theorem. In addition, a simple estimate of the variance of $\hat{m}(x)$ is available through expression (4.2) and an estimate of σ.

The only feature which causes difficulty is the bias present in the estimator $\hat{m}(x)$. If the value of this bias, denoted by $b(x)$, were known then a confidence interval could be constructed from the approximate pivotal result

$$\frac{\hat{m}(x) - m(x) - b(x)}{\sqrt{\hat{v}(x)}} \sim \mathcal{N}(0,1),$$

where $\hat{v}(x)$ denotes the estimated variance of $\hat{m}(x)$ and $\mathcal{N}(0,1)$ denotes the standard normal distribution. It is possible to substitute an estimate of $b(x)$ into the left hand side and continue to use this approximate result. However, since the expression for bias involves the second derivative of the true curve $m(x)$ this becomes a considerably more complex estimation problem. In addition, estimation of this term will increase the small sample variance beyond the simple expression $v(x)$. A thorough discussion of this approach, and practical recommendations, are given by Eubank and Speckman (1993). Other authors have sought to create confidence bands for the entire curve by computing the asymptotic distribution of the maximum deviation of the estimate from the true regression function. Another approach is to adopt a Bayesian formulation for the smoother. References on these approaches are given at the end of this chapter. Hastie and Tibshirani (1990, Section 3.8) give a general discussion on the problem of constructing confidence bands for nonparametric regression curves.

An alternative approach is to be content with indicating the level of variability involved in a nonparametric regression estimator, without attempting to adjust for the inevitable presence of bias. Bands of this type are easy to construct but require careful interpretation. Formally, the bands indicate pointwise confidence intervals for $\mathbb{E}\{\hat{m}(x)\}$ rather than $m(x)$. An estimate of the variance of $\hat{m}(x)$

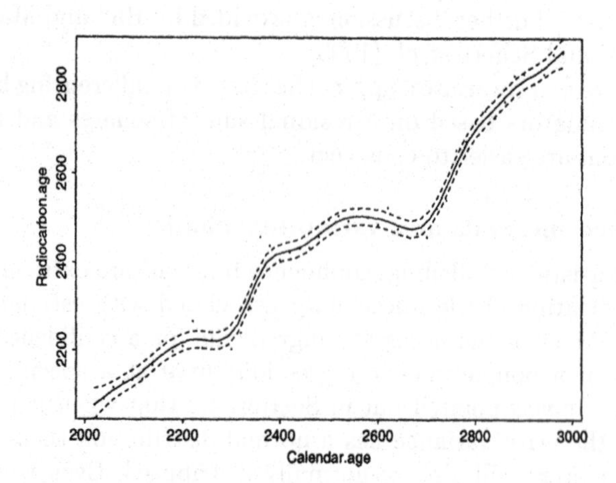

FIG. 4.2. Variability bands around the nonparametric regression estimate constructed from the radiocarbon data.

is all that is required. Bands can be constructed to indicate the size of two standard errors above and below $\hat{m}(x)$. In order to distinguish this from a proper confidence band, the term *variability band* will be used, as in the case of density estimation, discussed in Section 2.3.

Figure 4.2 displays a variability band around a nonparametric regression estimate constructed from the radiocarbon data. This is a helpful indication of the degree of variability present in the estimate. However, it cannot be used to draw firm conclusions about the shape of the curve in particular regions. This is not only because of the presence of bias, but also because of the pointwise nature of the bands. Formal methods of assessing the structure of some features of nonparametric regression curves are discussed in some detail in later sections of this book.

The concept of a variability band also applies to the nonparametric extensions of generalised linear models which were discussed in Section 3.4. At any given value of x, the likelihood calculations generate an estimated covariance matrix of the parameters. From this, a confidence interval for the predictor can be calculated, which in turn gives a confidence interval for the estimate $\hat{m}(x)$. In this way, assessments of the accuracy of the estimate can be computed locally, and combined to form a variability band. The bands displayed in Fig. 3.6 were produced by this method.

S-Plus Illustration 4.1. A variability band for the radiocarbon data

The following S-Plus *code may be used to reconstruct Fig. 4.2.*

```
provide.data(radioc)
Calendar.age     <- Cal.age[Cal.age>2000 & Cal.age<3000]
Radiocarbon.age  <-  Rc.age[Cal.age>2000 & Cal.age<3000]
```

```
sm.regression(Calendar.age, Radiocarbon.age, h = 30,
       display = "se")
```

4.5 Choice of smoothing parameters

In Section 4.2 the discussion of the simple properties of nonparametric regression estimators showed that bias increases with the size of the smoothing parameter h, while variance decreases. In order to define a suitable level of smoothing it is therefore necessary to find some compromise between these two tendencies. A simple strategy is to define at each point x the *mean squared error* $\mathbb{E}\{\hat{m}(x) - m(x)\}^2$, which is the sum of the squared bias and variance terms. To create a measure of performance which represents the global behaviour of the estimator, the mean squared errors can be summed over the observed design points. To prepare the way for asymptotic analysis, it is natural to consider the mean integrated squared error given by

$$\text{MISE}\,(h) = \int \mathbb{E}\{\hat{m}(x) - m(x)\}^2 f(x)\,dx,$$

where $f(x)$ represents the density of observed design points. The MISE is a function of the smoothing parameter h and so an optimal value h_{opt} can be defined as the value which minimises this quantity.

Use of the asymptotic expressions (4.3) and (4.4) in the evaluation of MISE leads to an optimal smoothing parameter which can be written as

$$h_{\text{opt}} = \left\{ \frac{\gamma(w)\sigma^2}{\int [m''(x)]^2 f(x) dx\, n} \right\}^{1/5}, \tag{4.8}$$

where $\gamma(w) = \alpha(w)/\sigma_w^4$. As expected, the optimal value of smoothing parameter depends on the unknown curve $m(x)$.

In density estimation the normal distribution provided a natural reference model from which a normal optimal smoothing parameter could be constructed. In the context of regression there is no such natural reference curve. For example, the optimal level of smoothing for a linear regression is infinite, since a local linear estimator approaches a linear regression as the smoothing parameter increases. Fan and Gijbels (1996, Section 4.2) describe an approach which involves polynomial approximations of the unknown quantities in (4.8).

Cross-validation has provided a popular means of selecting smoothing parameters by constructing an estimate of MISE and minimising this over h. The philosophy of cross-validation is to attempt to predict each response value y_i from the remainder of the data. For the value y_i, this predictor can be denoted by $\hat{m}_{-i}(x_i)$, where the subscript $-i$ denotes that the observation (x_i, y_i) has been omitted. The cross-validation function is then defined as

$$\text{CV}(h) = \frac{1}{n} \sum_{i=1}^{n} \{y_i - \hat{m}_{-i}(x_i)\}^2.$$

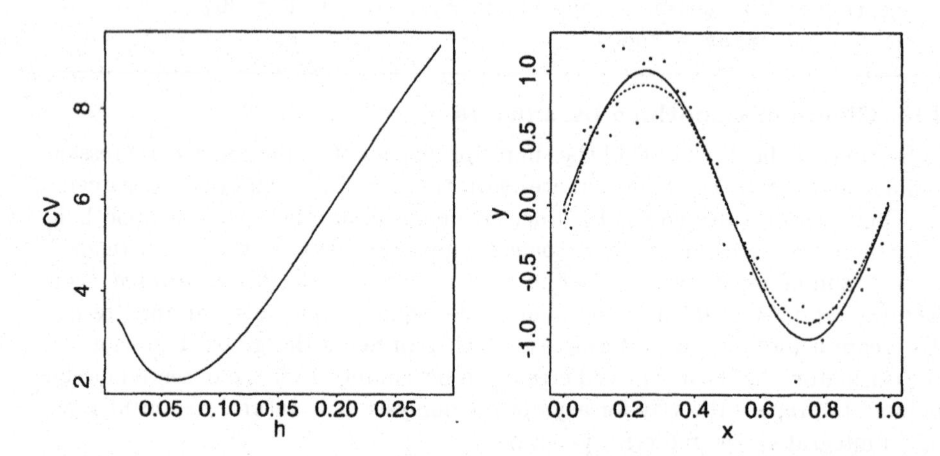

FIG. 4.3. The left panel shows the cross-validation curve with simulated data from a sine curve. The right panel shows the nonparametric regression curve (dotted) constructed from h_{cv} together with the true curve (full).

Some simple algebra shows that

$$\mathbb{E}\{\mathrm{CV}(h)\} = \frac{1}{n}\sum \mathbb{E}\{\hat{m}_{-i}(x_i) - m(x_i)\}^2 + \sigma^2.$$

The averaging over the design points x_i provides a discrete analogue of the integral and factor $f(x)$ in the MISE curve defined above, and so $\mathrm{CV}(h)$ provides a simple estimator for $\mathrm{MISE}(h)$, apart from the unimportant additive constant σ^2. Hastie and Tibshirani (1990, Section 3.4) give a helpful discussion of cross-validation and describe the modification known as *generalised cross-validation*, which has a simpler construction.

Figure 4.3 shows a cross-validation curve for data simulated from the sine function used in Section 4.2. The value of h which minimises this curve is $h_{cv} = 0.059$. This produces a nonparametric regression estimate which tracks the true sine function very closely.

An alternative approach is to construct a direct, or 'plug–in', estimator of the optimal smoothing parameter defined in (4.8). This is not an easy problem because it is necessary to estimate the second derivative of the regression function. Gasser *et al.* (1991) give expressions for the h_{opt} appropriate to the Gasser–Müller estimator and describe how the unknown quantities can be estimated effectively. Ruppert *et al.* (1995) have done this for the local linear estimator and Fan and Gijbels (1995) describe how a similar approach can be used for variable bandwidths. These techniques require more sophisticated estimation procedures but they provide a very promising means of selecting smoothing parameters.

As in the case of density estimation, one advantage of the cross-validatory approach is that the general nature of its definition allows it to be applied to a wide variety of settings. For example, the function $\mathrm{CV}(h)$ can easily be given an appropriate definition when the estimator $\hat{m}(x)$ is defined in terms of variable bandwidths, using nearest neighbour distances, such as the `lowess` curve of Cleveland(1979) or its `loess` equivalent.

The radiocarbon data of Section 3.2 provide an example of a setting where the principal interest lies in estimation of the underlying regression curve. The radiocarbon age of an object can be projected back through the curve to provide an estimate of its true calendar age. The role of the smoothing parameter in defining the estimate of the curve is therefore very important. However, there are many other regression problems where interest lies in particular features of the curve. For example, this might include whether the covariate has any effect on the response, whether this effect might be linear, and, in the case where effects are clearly nonparametric in nature, whether these effects are the same for different groups of data. Problems of this type form the principal focus of much of this book and it will be observed in later chapters that these inferential questions often shed a different light on the role of the smoothing parameter. From this point of view, strategies for selecting smoothing parameters based on ideas of mean integrated squared error are certainly of interest, but they are not necessarily of paramount importance.

S-Plus Illustration 4.2. Cross-validation to choose a smoothing parameter

---

The following S-Plus *code may be used to reconstruct Fig. 4.3.*

```
n <- 50
x <- seq(0, 1, length = n)
m <- sin(2 * pi * x)
h <- 0.05
sigma <- 0.2
y <- rnorm(n, m, sigma)
par(mfrow=c(1,2))
h.cv <- hcv(x, y, display="lines", ngrid=32)
plot(x, y)
lines(x, m)
sm.regression(x, y, h=hcv(x, y), add=T, lty=2)
par(mfrow=c(1,1))
```

4.6 Testing for no effect

In the exploration of catch score for the reef data, described in Chapter 3, longitude emerged as the most relevant explanatory variable. However, the possibility was also raised that this is largely an effect of depth, which is likely to increase broadly with distance from the shore. Figure 4.4 displays a plot of catch score against depth. The nonparametric regression curve superimposed on the plot is flat in the central region, with some indications of change at the ends of the depth range. However, there are very few points in the regions where the curve changes markedly, and some curvature in nonparametric estimates must always be expected, due to sampling variation. A means of identifying systematic pattern from natural variability would be extremely useful.

Variability bands are helpful in indicating the variation associated with nonparametric curves. However, in order to carry out a more formal assessment of shape, in a global manner, more traditional ideas of model comparison can be used. In the present example there are two competing models, namely

$$H_0 : \mathbb{E}\{y_i\} = \mu,$$
$$H_1 : \mathbb{E}\{y_i\} = m(x_i).$$

In linear models a standard approach is to compare the residual sums of squares from competing, nested models and to employ results on χ^2 and F distributions in a formal model comparison. In the nonparametric setting, residual sums of squares can easily be constructed as

$$\mathrm{RSS}_0 = \sum_{i=1}^{n} \{y_i - \bar{y}\}^2,$$
$$\mathrm{RSS}_1 = \sum_{i=1}^{n} \{y_i - \hat{m}(x_i)\}^2,$$

where the subscripts 0 and 1 refer to the corresponding hypotheses. Following the linear model analogy, a suitable test statistic is proportional to

$$F = \frac{(\mathrm{RSS}_0 - \mathrm{RSS}_1)/(\mathrm{df}_0 - \mathrm{df}_1)}{\mathrm{RSS}_1/\mathrm{df}_1},$$

where df_0 and df_1 denote the degrees of freedom for error, mentioned in Section 4.3, under each hypothesis. This has the particular advantage of being independent of the unknown error variance parameter σ^2, as a result of the ratio form. However, the necessary conditions for each residual sum of squares to have a χ^2 distribution, and for their scaled comparison to have an F distribution, do not apply in the context of nonparametric regression.

A method of calculating the distribution of F in an accurate way will be described in Chapter 5. For the moment, a more computationally intensive method will be adopted, based on the idea of a permutation test described by Raz (1990). This rests on the fact that when H_0 is true the pairing of any particular x and y in the observed sample is entirely random. The distribution of the F statistic defined above can therefore be generated by simulation, using random pairings of the observed xs and ys and constructing the corresponding F statistic in each case. The empirical p-value of the test is then simply the proportion of

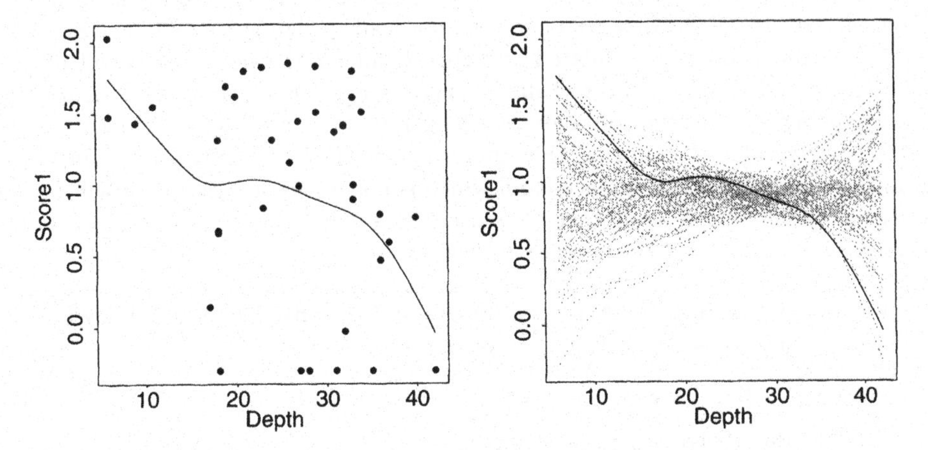

FIG. 4.4. A plot of the relationship between catch score and depth (left) and a reference band for the no-effect model.

simulated F statistics which are larger than the one observed from the original data. For the effect of depth on the catch score in the reef data the empirical p-value was found to be 0.04. This suggests that there is a relationship between the two variables. However, it is clear that longitude has a much stronger effect on catch score than depth, and since longitude and depth are themselves closely related, this points towards the retention of longitude as the principal covariate in the model. This is consistent with the explanation that the catch score is influenced by the nature of the sea bed, which changes with distance from the shore.

Readers are invited to explore whether there is an additional effect of depth, after adjusting for longitude, in an exercise in Chapter 8.

4.7 A reference band for the no-effect model

The model comparison described above provides a useful global assessment of whether a covariate has any effect on the mean value of a response. Variability bands can be used in this context to provide a helpful graphical follow-up. However, an alternative approach is available through the random permutations which are generated in the test procedure described in the previous section. Each set of randomly permuted data can be smoothed to produce a nonparametric regression curve. Together, these curves indicate where a nonparametric regression curve is likely to lie when the hypothesis of no relationship between catch score and depth is correct. This collection of curves forms an analogue of the *reference band* described for a normal model in density estimation in Chapter 2, as it indicates where a curve will lie under the reference model. This can be helpful in indicating what features of the observed curve are associated with departures from the reference model, or in explaining, through the variance structure, why apparent features of the observed curve do not lead to significant differences.

Figure 4.4 displays a simulated reference band for the no-effect model for catch score and depth in the reef data. There is again evidence, although not overwhelming, of a relationship between the variables, as the shape and location of the curve produced from the original data are at the limits of those produced from the simulations. A reference band which is not based on simulation will be described in Chapter 5, and other reference bands will be used in later chapters to explore a variety of other regression hypotheses.

S-Plus Illustration 4.3. A reference band for a no-effect model with the reef data, constructed by permutations.

The following S-Plus *code may be used to reconstruct Fig. 4.4.*

```
provide.data(trawl)
ind <- (Year == 1 & Zone == 1 & !is.na(Depth))
par(mfrow=c(1,2))
sm.regression(Depth[ind], Score1[ind], h = 5,
      xlab="Depth", ylab="Score1")
plot(Depth[ind], Score1[ind], type = "n",
      xlab="Depth", ylab="Score1")
for (i in 1:100)
   sm.regression(Depth[ind], sample(Score1[ind]),
   h = 5, col = 6, lty = 2, add = T)
sm.regression(Depth[ind], Score1[ind], h = 5, add = T)
par(mfrow=c(1,1))
```

4.8 The bootstrap and nonparametric regression

The bootstrap can often provide a very useful means of deriving the properties of estimators, and of constructing confidence intervals, when theoretical analysis proves difficult to pursue. This is an attractive possibility in the case of nonparametric regression, where the form of the bias in particular introduces complications, as discussed in Section 4.4. In order to apply the bootstrap in the context of nonparametric regression, Härdle and Bowman (1988) proposed the following algorithm. For convenience, the notation $\hat{m}(x;h)$ is adopted for the estimator in order to make the values of the smoothing parameter explicit.

1. Construct residuals $\hat{\varepsilon}_i = y_i - \hat{m}(x_i;h_p)$ through a pilot estimator $\hat{m}(x;h_p)$.
2. Create a set of normalised residuals $\tilde{\varepsilon}_i = \hat{\varepsilon}_i - \frac{1}{n}\sum_j \hat{\varepsilon}_j$, with mean 0.
3. Repeatedly create bootstrap observations $y_i^* = \hat{m}(x_i;h_p) + \varepsilon_i^*$, through the pilot estimator $\hat{m}(x_i;h_p)$ and random sampling of ε_i^* from $\{\tilde{\varepsilon}_j\}$.
4. Repeatedly create bootstrap estimators $\hat{m}^*(x;h)$ by smoothing the observations (x_i, y_i^*).

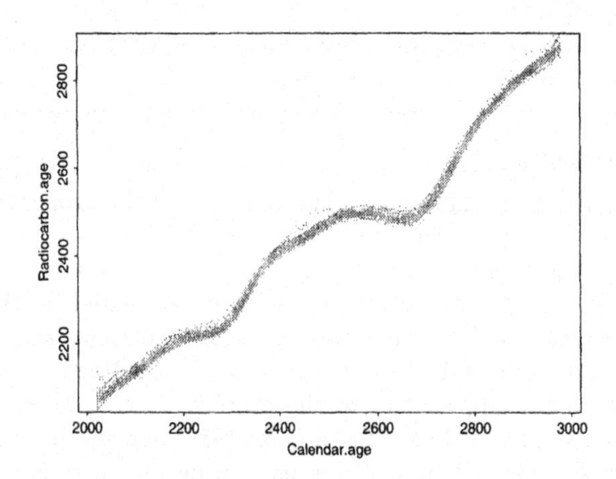

FIG. 4.5. Bootstrap regression curves for the radiocarbon data.

Härdle and Bowman (1988) and Härdle and Marron (1991) analysed this algorithm for simple estimators $\hat{m}$ and reached the disappointing conclusion that for the distribution of $\hat{m}^*$ about $\hat{m}$ to mimic effectively the distribution of $\hat{m}$ about m it is necessary either to estimate the bias of $\hat{m}$ explicitly or to use a pilot smoothing parameter h_p which is not of the magnitude required for optimal performance of an estimator. Once again, it the bias which is the source of the problem.

Since the bootstrap does capture the variability of the estimate effectively, it can therefore be used as an alternative means of generating variability bands, as discussed in Section 4.4. Figure 4.5 displays a band of this type for the radiocarbon data. This is very similar to the band shown in Fig. 4.2. The bootstrap should have some advantage in cases where the distribution of the data about the true regression line is skewed, as the resampling process should capture some of the asymmetric effects of this.

S-Plus Illustration 4.4. A variability band produced by the bootstrap.

The following S-Plus *code may be used to reconstruct Fig. 4.5.*

```
provide.data(radioc)
x <- Cal.age[Cal.age>2000 & Cal.age<3000]
y <-  Rc.age[Cal.age>2000 & Cal.age<3000]
plot(x, y, xlab="Calendar.age", ylab="Radiocarbon.age",
        type="n")
model <- sm.regression(x, y, h=30, eval.points=x,
        display="none")
mhat <- model$estimate
r    <- y - mhat
```

```
r     <- r - mean(r)
for (i in 1:50) sm.regression(x, mhat + sample(r, replace=T),
        h=30, add=T, col=6, lty=2)
```

4.9 Further reading

Fan and Gijbels (1995) provide an excellent overview of the properties and behaviour of local polynomial estimators. Green and Silverman (1994) perform a similar service for the spline approach.

Confidence intervals for an entire regression curve, based on the maximum deviation of the estimate from the true regression function, are discussed by Knafl *et al.* (1985), Hall and Titterington (1988) and Härdle (1989), among other authors. Bayesian approaches to the problem of constructing confidence intervals are discussed by Wahba (1983), Cox (1986) and Nychka (1988; 1990).

Fan and Gijbels (1996, Chapter 4) discuss a wide variety of issues associated with bandwidth choice in nonparametric regression. Theoretical aspects in particular are discussed, for example, by Härdle *et al.* (1988) and Hall and Johnstone (1992).

The difficult problem of choosing bandwidths which adapt to the local curvature of the regression function is addressed by Brockmann *et al.* (1993).

Exercises

4.1 *Means and standard deviations.* A script to reproduce Fig. 4.1 is available. Use this to explore the effect of changing the values of `n`, `h` and `sigma`. Also compare the properties of the local linear estimator with those of the local constant approach by setting `poly.index = 0`. Finally, explore the effect of using a `sin` function with a higher frequency.

4.2 *Accuracy of asymptotic expressions for mean and standard deviations.* Amend the script for Fig. 4.1 to produce means and standard deviation bands based on the asymptotic expressions given in Section 4.2. Explore the adequacy of these approximations by comparing them with the exact expressions, for a variety of parameter and function settings.

4.3 *Estimation of σ^2.* Use S-Plus to construct a graph for the radiocarbon data which displays an estimator of the error variance σ^2 based on residual sums of squares and approximate degrees of freedom as a function of the smoothing parameter h. The weight matrix S can be constructed through the function `sm.weight(x, y, h)`. Add to the plot a horizontal line corresponding to the difference based estimator of Gasser *et al.* (1986), which is available in the function `sm.sigma(x, y)`.
To what value will the residual sum of squares based estimator converge as the smoothing parameter becomes very large?

4.4 *Variability bands for the radiocarbon data.* Construct variability bands for the radiocarbon data, as in Fig. 4.5, and superimpose variability bands as displayed in Fig. 4.2. Compare the two sets of bands.

4.5 *Assessing a regression effect with survival data.* Employ the permutation method described in Sections 4.6 and 4.7 to produce simulated curves which indicate where an estimate of the median curve is likely to lie when log survival time and age are unrelated. Use this graph to assess visually whether there is a relationship between the two variables.

4.6 *Variability bands for the low birthweight data.* Plot variability bands for the two curves derived from the low birthweight data by setting `display="se"` in `sm.logit`. Use these bands to help interpret whether there are significant differences between the two curves at particular points.

4.7 *Variability bands for the worm data.* Construct variability bands for the curves describing the probability of occurrence of the parasitic infection as a function of age, for the male and female groups in the worm data. Use these bands to help identify the nature of changes over time and to compare the two sex groups.

5

CHECKING PARAMETRIC REGRESSION MODELS

5.1 Introduction

In Chapters 1 and 3, smoothing techniques were introduced as descriptive and graphical tools for exploratory data analysis. A subsequent step in many cases will be the formulation of a parametric model which takes into account the features which have emerged from the preliminary analysis. The important process of formulating and fitting a suitable parametric model is outside the scope of this book.

However, there will always be specific questions which are to be asked of the data. There may indeed be associated theory or hypotheses which have implications for an appropriate statistical model, even at a very simple level such as assumed linear relationships. In these circumstances, such models clearly take precedence over an entirely speculative, exploratory approach. This in turn requires a different strategy for statistical analysis.

The purpose of the present chapter is therefore to investigate the role of nonparametric smoothing techniques in providing helpful tools for checking the appropriateness of a proposed parametric model. Attention is devoted to regression models and to generalised linear models. However, for the purposes of exposition, a particularly simple case is considered first.

5.2 Testing for no effect

In Section 4.6, the relationship between depth and catch score was investigated for the reef data. A plot of the data, displayed in Fig. 4.4, does not immediately settle the question of whether the mean catch score changes with depth. As a means of separating systematic pattern from natural variability, the analysis of Section 4.6 used a permutation test to examine the evidence that these two variables are related.

In order to discuss an alternative means of inference it is helpful to repeat the formal structure of the problem. There are two competing models, namely

$$\begin{aligned} H_0 &: \mathbb{E}\{y_i\} = \mu, \\ H_1 &: \mathbb{E}\{y_i\} = m(x_i). \end{aligned}$$

The residual sums of squares offer a natural means of quantifying the extent to which these models explain the data. These are

$$\mathrm{RSS}_0 = \sum_{i=1}^{n} \{y_i - \bar{y}\}^2,$$

$$\text{RSS}_1 = \sum_{i=1}^{n} \{y_i - \hat{m}(x_i)\}^2,$$

where the subscripts indicate the hypothesis under which each is calculated. A means of quantifying the difference between the residual sums of squares is provided by a statistic such as

$$F = \frac{\text{RSS}_0 - \text{RSS}_1}{\text{RSS}_1}$$

which is proportional to the usual F statistic and whose construction as a ratio scales out the effect of the error variance σ^2.

A suitable name for F is the *pseudo-likelihood ratio test* (PLRT) statistic. This indicates the formal analogy with a traditional likelihood ratio, but the prefix 'pseudo' acts as a reminder that the current situation lies outside the traditional setting. In particular, the 'alternative hypothesis' is not a parametric class of models and the estimate is not fitted by maximum likelihood.

In order to proceed further, it is necessary to find the distribution of the statistic F under the model H_0. The following discussion is based on the assumption that the errors ε have a normal distribution. Where this is inappropriate the permutation test of Section 4.6 can be applied. For the reef data an assumption of normality is reasonable, with the exception of a few observations where the detection limit has been reached at the lower end of the catch score scale. Leaving this aside for the moment, a normal model will be adopted as a working approximation.

It is helpful to express the structure of F in terms of quadratic forms. The two residual sums of squares can be written as

$$\begin{aligned}
\text{RSS}_0 &= y^\top (I - L)^\top (I - L) y = y^\top (I - L) y, \\
\text{RSS}_1 &= y^\top (I - S)^\top (I - S) y,
\end{aligned}$$

where L denotes an $n \times n$ matrix filled with the value $1/n$, and S is the smoothing matrix used to create the vector of fitted values $\{\hat{m}(x_i)\}$. The statistic can then be expressed as

$$F = \frac{y^\top B y}{y^\top A y},$$

where A is the matrix $(I-S)^\top (I-S)$ and B is the matrix $I-L-A$. Unfortunately, standard results from linear models do not apply because the matrices A and B do not have the necessary properties, such as positive definiteness.

Fortunately, there are results which can be used to handle statistics of this kind under more general conditions. These results require only that the matrices from which the quadratic forms are created are symmetric, which is the case here. As a first step it is helpful to reformulate the problem, and to focus on the significance of the statistic F as expressed in its p-value. This is

$$p = \mathbb{P}\{F > F_{\text{obs}}\},$$

where F_{obs} denotes the value calculated from the observed data. This can be re-expressed as

$$\begin{aligned} p &= \mathbb{P}\left\{\frac{y^\top By}{y^\top Ay} > F_{\text{obs}}\right\} \\ &= \mathbb{P}\{y^\top Cy > 0\}, \end{aligned}$$

where

$$C = (B - F_{\text{obs}}A).$$

Johnson and Kotz (1972, pp. 150–153) summarise general results about the distribution of a quadratic form in normal variables, such as $y^\top Cy$, for any symmetric matrix C. These results can be applied most easily when the normal random variables have mean zero. In the present setting, y_i has mean μ under the null hypothesis. However, it is easy to see from the form of the residual sums of squares RSS_0 and RSS_1, which are the building blocks of the test statistic F, that μ disappears from the scene because of the differences involved. The quadratic form $y^\top Cy$ is therefore equivalent to the quadratic form $Q = \varepsilon^\top C\varepsilon$. The results of Johnson and Kotz (1972) then allow the probability p defined above to be calculated exactly, although in numerical rather than analytical form.

However, a highly accurate computation is unnecessary, since two significant digits are generally sufficient for a p-value. In addition, the full calculations become awkward when n is large. It is sufficient instead to use an approximate evaluation of p, by replacing the real distribution of Q by another more convenient distribution with the same first three or four moments. This approach is known to work well in a number of similar cases, and it is particularly suited to the present calculations since the cumulants of Q have a simple explicit expression, namely

$$\kappa_j = 2^{j-1}(j-1)!\,\text{tr}\{(VC)^j\},$$

where $\text{tr}\{\cdot\}$ denotes the trace operator, and V is the covariance matrix of y. Here V is simply the identity matrix, but it is useful to state the general formula for later reference.

In many problems involving quadratic forms, a shifted and scaled χ^2 distribution has often been found to provide a very good approximation. By matching the moments of an $a\chi^2_b + c$ distribution with the moments of Q, we find that

$$a = |\kappa_3|/(4\kappa_2), \quad b = (8\,\kappa_2^3)/\kappa_3^2, \quad c = \kappa_1 - a\,b. \tag{5.1}$$

The p-value of interest can then be accurately approximated as $1 - q$, where q is the probability lying below the point $-c/a$ in a χ^2 distribution with b degrees of freedom.

An application of this procedure to catch score and depth in the reef data produces a p-value of 0.06, using a smoothing parameter of 5, which corresponds to the nonparametric regression curve displayed in Fig. 4.4. This p-value

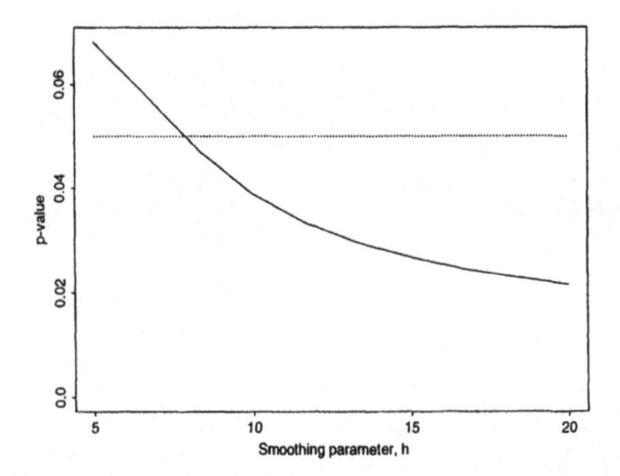

FIG. 5.1. A significance trace to assess the evidence for a relationship between catch score and depth in the Great Barrier Reef data.

is very close to the one produced by the permutation test and it is sufficiently small to lend further credence to the suggestion that catch score and depth are related. However, the analysis can be repeated for a wide variety of other values of smoothing parameter. In order to reflect this, Fig. 5.1 plots these p-values as a function of the smoothing parameter h. This is referred to as a *significance trace*. From this plot we can see that the evidence for a relationship between catch score and depth is similar across a wide range of values of h. This can be a helpful way of expressing the evidence in the data, as it avoids the need to settle on a particular degree of smoothing to apply.

The local linear technique has been used here. This has the consequence that as the smoothing parameter reaches very large values the estimate will approach the least squares regression line of log weight on depth. It also means that the significance of the nonparametric regression curve will approach that of simple linear regression, which for these data has a p-value of 0.02. The nonparametric analysis therefore allows the linear assumption to be relaxed, with the smoothing parameter controlling the extent of this relaxation.

The formal model comparison described above provides a useful global assessment of whether a covariate has any effect on the mean value of a response. Variability bands can be used in this context to provide a helpful graphical follow-up. The reference model of interest is $H_0 : \mathbb{E}\{y_i\} = \mu$. Under this model a nonparametric estimate $\hat{m}(x)$, represented as $\sum v_i y_i$, has no bias, since $\mathbb{E}\{\sum v_i y_i\} = \sum v_i \mu = \mu$. A *reference band* can therefore be placed around this model, to indicate where the nonparametric regression curve should lie, under the null hypothesis. The variance of the difference between the nonparametric model and the reference model is

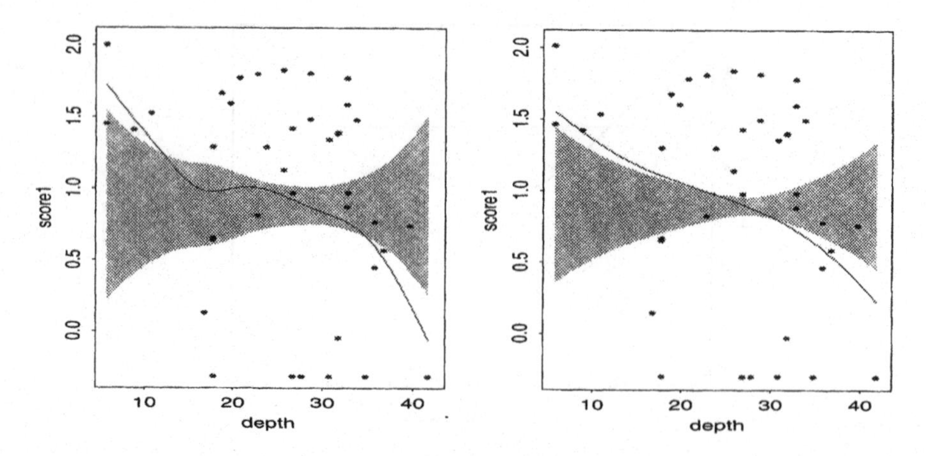

FIG. 5.2. Reference bands for the 'no-effect' model for the species weight and depth relationship in the Great Barrier Reef data. A smoothing parameter of 5 was used in the left panel, and 10 in the right panel.

$$\text{var}\left\{\sum_i v_i y_i - \frac{1}{n}\sum_i y_i\right\} = \sum_i \left(v_i - \frac{1}{n}\right)^2 \sigma^2.$$

The error variance σ^2 can be estimated by the differencing method described in Section 4.3. At any point x, the reference band stretches from $\bar{y} - 2\hat{\sigma}b$ to $\bar{y} + 2\hat{\sigma}b$, where b denotes the quantity $\sqrt{\sum_i (v_i - 1/n)^2}$. (Note that the dependence of the value of v_i on x has been suppressed in this notation, for simplicity.) If the curve exceeds the band then this indicates that the parametric and nonparametric models are more than two standard errors apart, under the assumption that the parametric model is correct. This can be helpful in indicating what features of the observed curve are associated with departures from the reference model, or in explaining, through the variance structure, why apparent features of the observed curve do not lead to significant differences.

Figure 5.2 displays reference bands for the relationship between catch score and depth in the reef data. Two different values of smoothing parameter have been used. In both cases the curve exceeds the band at each end of the depth range, confirming that there is some evidence of a relationship between the two variables.

S-Plus Illustration 5.1. A test of significance for the relationship between weight and latitude for the Great Barrier Reef data

Figure 5.1 was constructed with the following S-Plus *code.*

```
provide.data(trawl)
ind        <- (Year == 1 & Zone == 1 & !is.na(Depth))
score1     <- Score1[ind]
```

```
depth     <- Depth[ind]
summary(lm(score1 ~ depth))
sig.trace(sm.regression(depth, score1,
        model = "no.effect", display="none"),
        hvec = seq(5, 20, length = 10))
```

S-Plus Illustration 5.2. Reference bands for the no-effect model in the Great Barrier Reef data.

Figure 5.2 was constructed with the following S-Plus *code.*

```
provide.data(trawl)
ind       <- (Year == 1 & Zone == 1 & !is.na(Depth))
score1    <- Score1[ind]
depth     <- Depth[ind]
par(mfrow=c(1,2))
sm.regression(depth, score1, h = 5, model = "no.effect")
sm.regression(depth, score1, h = 10, model = "no.effect")
par(mfrow=c(1,1))
```

5.3 Checking a linear relationship

For the purposes of exposition, the *pseudo-likelihood ratio test* was described in Section 5.2 in a particularly simple situation. However, this approach can be adapted to deal with a variety of other models. An example is provided by examining the relationship between catch score and longitude for the closed zone in 1993 with the reef data, as displayed in the left panel of Fig. 5.3. A natural starting point for a parametric model is a simple linear regression,

$$y = \alpha + \beta x + \varepsilon.$$

This has been fitted by least squares and superimposed on the plot of the data. The earlier, exploratory examination of catch score and longitude suggested a 'shelf-like' pattern. By examining the suitability of a simple linear model an assessment can be made of the strength of the evidence for a nonlinear relationship.

Two standard tools for the problem are the use of graphical diagnostics and the insertion of a quadratic term in the linear model, followed by a test on the significance of the associated coefficient. The former approach would lead to a plot similar to the right panel of Fig. 5.3, which shows the scatterplot of the residuals

$$e_i = y_i - \hat{\alpha} - \hat{\beta} x_i$$

against longitude. This plot does not indicate any striking departure from linearity. The quadratic approach leads to an observed significance level for the coefficient of x^2 equal to 0.46. Here again there is no indication of inadequacy of the fitted linear model.

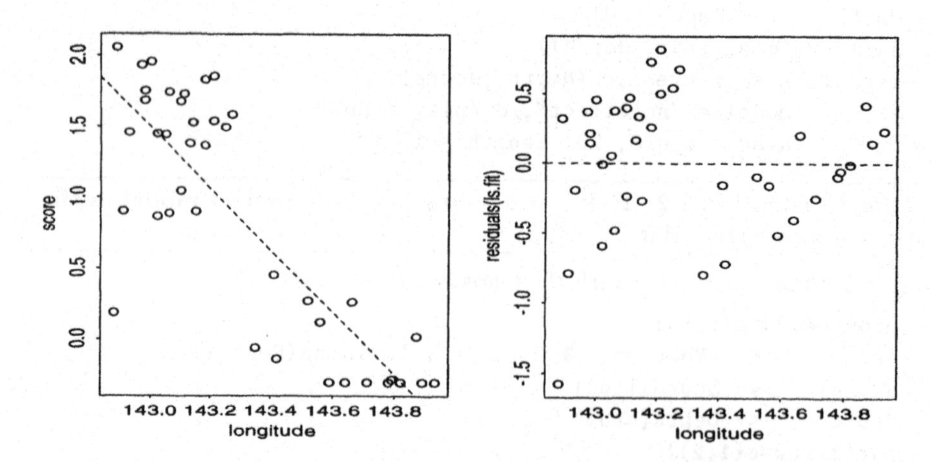

FIG. 5.3. A linear model fitted to the Great Barrier Reef data for the closed zone in 1993. The left panel shows the scatterplot of catch score against longitude, with a linear regression superimposed. The right panel plots the residuals against longitude.

An alternative approach is to extend the argument of Section 5.2, again making use of the statistic

$$F = \frac{\text{RSS}_0 - \text{RSS}_1}{\text{RSS}_1},$$

except that the 'data' are now represented by the residual vector from the fitted model, namely $e = (e_1, \ldots, e_n)^\top$. The formal hypotheses test can be written as

$$H_0 : \mathbb{E}\{e\} = 0,$$
$$H_1 : \mathbb{E}\{e\} = \text{a smooth function of } x.$$

Since the expected value is 0 under the null hypothesis, RSS_0 is simply $\sum_i e_i^2$. Hence the new form of the F statistic is

$$F = \frac{e^\top Be}{e^\top Ae}$$

where $A = (I - S)^\top (I - S)$ as before, and $B = I - A$.

To compute the observed significance level p, the method described in Section 5.2 is still applicable. The p-value can be written as

$$\begin{aligned} p &= \mathbb{P}\{F > F_{\text{obs}}\} \\ &= \mathbb{P}\{e^\top (I - (1 + F_{\text{obs}})(I - S)^\top (I - S))e > 0\} \\ &= \mathbb{P}\{e^\top Ce > 0\}, \end{aligned}$$

which again leads to the computation of the distribution of a quadratic form in normal random variables with mean zero.

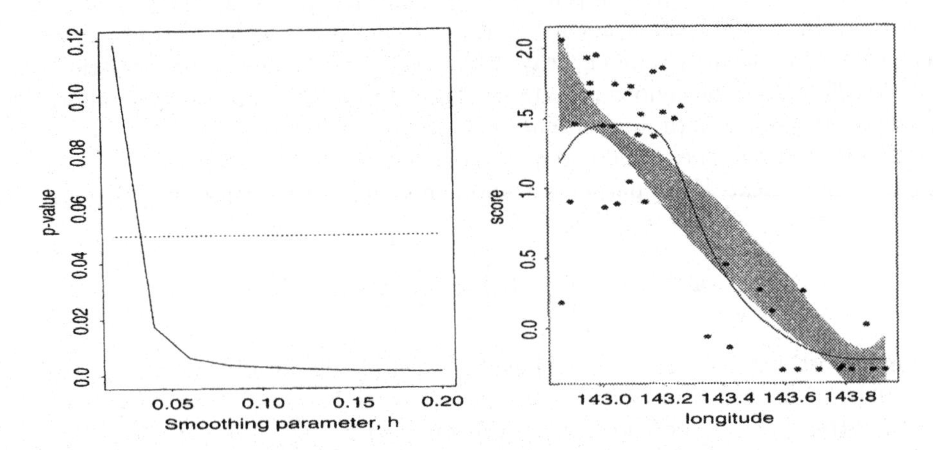

FIG. 5.4. Plots to assess linearity in the Great Barrier Reef data for the closed zone in 1993. The left panel provides the significance trace of the PLRT. The right panel shows reference bands for the $h = 0.1$

The principal difference from the case of Section 5.2 is that the components of e are correlated, since

$$\mathrm{var}\{e\} = I - X(X^\top X)^{-1}X^\top = V, \tag{5.2}$$

say, where X denotes the design matrix of the fitted linear model. Apart from this difference, the computational methods used earlier still apply, with an approximation obtained by matching the first three moments of the quadratic form with those of an $a\chi_b^2 + c$ distribution, with parameters given by (5.1)

Plotting the p-value against the smoothing parameter h again gives a *significance trace*, as described in Section 5.2. For the present data, the curve of Fig. 5.4 provides a strong indication that a simple linear model is inappropriate. This conclusion applies over a wide range of values of h.

This stability of the observed significance level with respect to the value of h, as exhibited in Figs 5.1 and 5.4, has been observed in a number of other datasets, to varying degrees. This behaviour underlines the difference between inference, where the aim is to detect the presence of some effect of interest by comparing models, and estimation, where interest is focused on the best possible estimate, in this case of the regression relationship. Some of the values of h used to produce the significance trace in Fig. 5.4 may result in curves which do not estimate the true regression relationship between catch score and longitude very effectively. However, almost all values of h are effective in identifying that this relationship is not linear.

The statistic F has an analogy with the Durbin–Watson statistic for detecting serial correlation in the error terms of a linear regression model. After a monotonic transformation, the principal difference lies in the quantity denoted here by RSS_1. The Durbin–Watson statistic handles the variability between adjacent values in a slightly different way. Although the Durbin–Watson statistic is rather different, its behaviour is quite similar to that of F in that they are both sensitive to slow, smooth fluctuations of the residuals.

The idea of reference bands presented in Section 5.2 extends immediately to the present context. If the parametric and nonparametric estimates at the point x are denoted by $\hat{\alpha} + \hat{\beta}x = \sum l_i y_i$ and $\hat{m}(x) = \sum v_i y_i$ respectively, then

$$\text{var}\left\{\hat{m}(x) - \hat{\alpha} - \hat{\beta}x\right\} = \sum_i (v_i - l_i)^2 \sigma^2.$$

A simple estimate of the error variance σ^2 is available by differencing of the residuals, as described in Section 4.3. The reference band can then be centred on the fitted linear model, extending a distance of $2\hat{\sigma}\sqrt{\sum(v_i - l_i)^2}$ above and below.

The right panel of Fig. 5.4 illustrates this for the choice $h = 0.1$. It is clear that the plateau in catch score for points near the coast (low longitude), and the subsequent rapid fall, are systematic features of the data which a simple linear model does not adequately describe. If a different value of h is used, the width of the band and the detailed shape of the curve change. However, this does not markedly change the indication of where, and to what extent, the reference bands are exceeded by the nonparametric estimate.

The methods described above can be adapted to other situations. For example, the x variable against which the residuals are plotted and smoothed does not need to be the same covariate which appears in the fitted linear model. In this case, the method can be used to test whether an additional variable should be added to the present model.

The associated distribution theory for the resulting test statistic follows as before. Note that the design matrix X which leads to the covariance matrix V of (5.2) remains the one used to obtain the initial least squares fit. Illustration 5.4 provides an example of this case. Here, the design matrix is formed from columns containing the constant 1 and the covariates for treatment and block effects, and x represents the serial position of the observations.

The use of the pseudo-likelihood ratio test and its associated significance trace is not limited to models which involve a single covariate x. Two or more covariates can be considered, although the associated graphical display is necessarily limited to the case of two covariates. Illustration 5.5 provides an example of this.

S-Plus Illustration 5.3. Checking a linear relationship

Figure 5.4 was constructed with the following S-Plus *code.*

```
provide.data(trawl)
ind       <- (Year == 0 & Zone == 1)
score     <- Score1[ind]
```

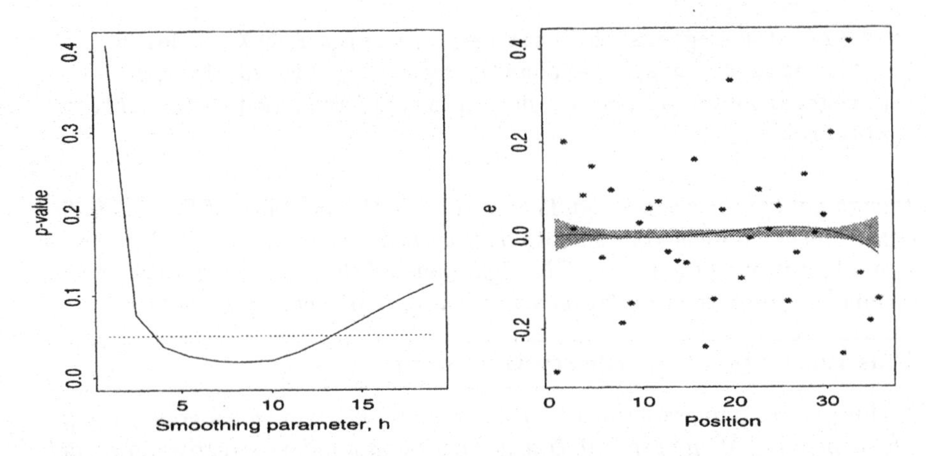

FIG. 5.5. A significance trace for the mildew data, to test the residual effect of serial position of the observations (left panel), and a reference band using $h = 7$ (right panel).

```
longitude <- Longitude[ind]
par(mfrow=c(1,2))
sig.trace(sm.regression(longitude, score,
        model = "linear", display="none"),
        hvec = seq(0.02, 0.2, length = 10))
sm.regression(longitude, score, h = 0.1, model = "linear")
par(mfrow=c(1,1))
print(summary(lm(score ~ poly(longitude,2))))
```

S-Plus Illustration 5.4. Mildew control data

Data on mildew control were presented by Draper and Guttman (1980) and subsequently analysed by a number of other authors, including Green et al. (1985). Four treatments for mildew control were compared in a single column of 38 plots, arranged in nine blocks of four plots, with an extra plot at each end. After a linear model which allows for treatment and block effects has been fitted to the data, it is of interest to investigate the presence of spatial trend, represented by a smooth residual fluctuation associated with the serial position of each plot in the column. In this case, x represents the serial position of the blocks.

An analysis is implemented by the following code, whose graphical output is presented in Fig. 5.5.

```
provide.data(mildew)
X <- cbind(rep(1,36),as.matrix(mildew[1:11]))
e <- residuals(lsfit(X,Yield,intercept=F))
Position <- 1:36
```

```
par(mfrow=c(1,2))
sig.trace(sm.regression(Position, e, design.mat=X, model =
    "no.effect", display="none"), hvec = seq(1, 20, by=1.5))
sm.regression(Position, e, design.mat=X, h=7, model="no.effect")
par(mfrow=c(1,1))
```

Although not conclusive, the significance trace in the left panel of Fig. 5.5 indicates the possible presence of a smooth fluctuation in the residuals associated with the serial position of the blocks. The right panel of the same figure shows where the departure from the simple analysis of variance model may be occurring.

S-Plus Illustration 5.5. Cherry trees data

The cherry trees data, described by Ryan et al. (1985), provide a good example of the benefits of an appropriate transformation in building a regression model. Since these data have been analysed by a large number of authors, the purpose here is not to search for new features of the data, but rather to use them to illustrate the methodology described in this section.

The data refer to the volume of usable timber from a sample of trees, together with the explanatory variables Diameter and Height. The model

$$\log(\textit{Volume}) = \textit{constant} + \log(\textit{Diameter}) + \log(\textit{Height})$$

enjoys both a simple geometric interpretation and a good empirical fit to the data.

The S-Plus *code below computes the pseudo-likelihood ratio test to examine linearity in four cases, namely:*

- *Volume versus Diameter, with observed significance* $p = 0.007$;
- *Volume versus Height, with observed significance* $p = 0.734$;
- *Volume versus (Diameter, Height), with observed significance* $p = 0.023$;
- *log(Volume) versus (log(Diameter), log(Height)), with observed significance* $p = 0.683$.

For simplicity, the values of the two smoothing parameters have been kept fixed at $h_1 = 1.5$ *and* $h_2 = 4$ *for Diameter and Height, respectively. These outcomes confirm the general findings of the literature.*

```
provide.data(trees)
h1 <- 1.5
h2 <- 4
par(mfrow=c(2,2))
par(cex=0.7)
sm.regression(Diameter, Volume, h= h1, model = "linear")
sm.regression(Height, Volume, h= h2, model = "linear")
X <- cbind(Diameter, Height)
sm.regression(X, Volume, h = c(h1,h2), model = "linear",
```

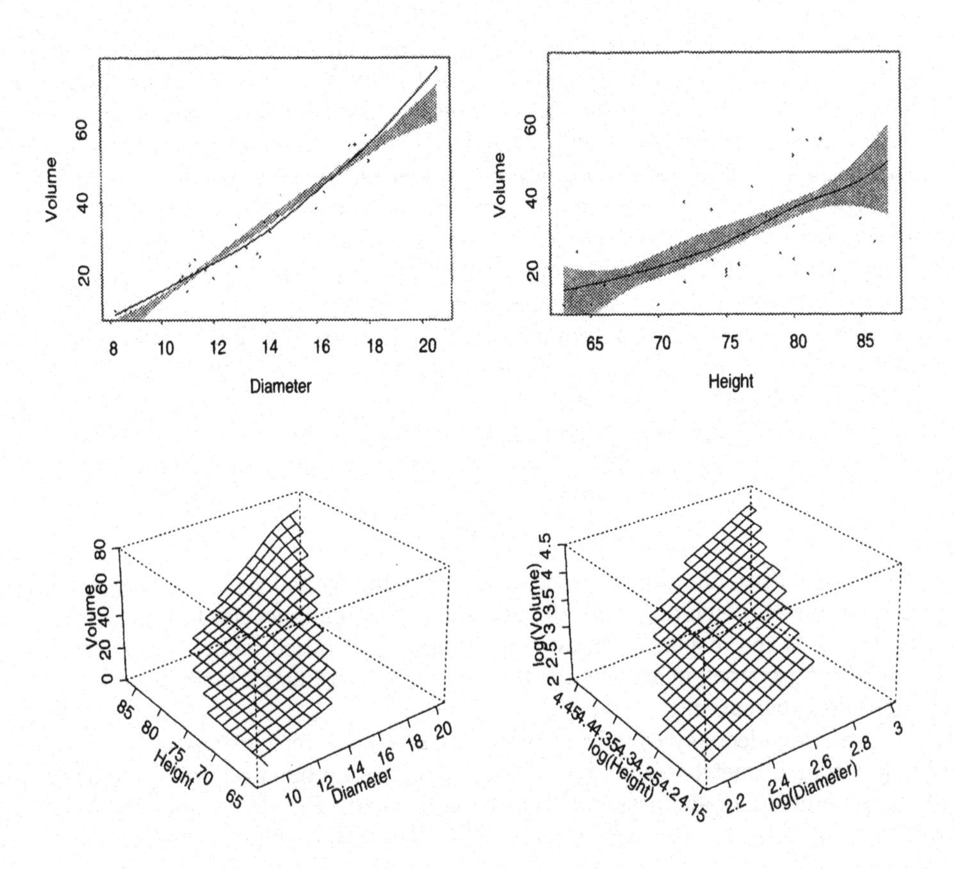

FIG. 5.6. Four different models from the cherry trees data.

```
    zlim = c(0,80))
X <- cbind(log(Diameter), log(Height))
dimnames(X)[[2]] <- c("log(Diameter)","log(Height)")
sm.regression(X, log(Volume), h = log(c(h1,h2)),
            model = "linear")
par(cex=1)
par(mfrow=c(1,1))
```

5.4 The pseudo-likelihood ratio test

In the previous sections, the reference parametric model was linear and the error terms were assumed to be normally distributed. This framework allows explicit quadratic forms for the test statistic to be obtained, with a correspondingly simple treatment of the required distribution theory.

This approach can be extended to examine the validity of parametric models in more complex situations. Beyond the well behaved context of normal linear models a simple, exact treatment of the necessary distribution theory is no longer possible. However, the underlying rationale remains valid, and an approximate test is feasible, as demonstrated in the remainder of this section.

A broad extension is provided by generalised linear models (GLMs), which involve a linear predictor and a link function in the specification of a parametric form $m(x;\theta)$ for the regression curve $m(x) = \mathbb{E}\{Y|x\}$. The discussion below is focused primarily on GLMs but extends to more general situations, provided a parametric form of regression function has been specified. The following discussion assumes some basic knowledge of GLMs. Dobson (1990) provides an excellent introductory account of this topic.

In a general setting, the reference parametric model gives the response variable y_i a distribution depending on a covariate value x_i through the relationship

$$y_i|x_i \sim f(\cdot; m(x_i;\theta), \psi),$$

where f is a density function from the exponential family and ψ denotes any additional parameter which does not influence the regression function. In the context of GLMs, ψ is the dispersion parameter. Under a proposed model, estimates $\hat{\theta}$ and $\hat{\psi}$ of the parameters are obtained, leading to a parametric estimate $m(x;\hat{\theta})$ of the regression curve.

To test the adequacy of the parametric model, two steps are involved: (i) to obtain a nonparametric estimate $\hat{m}(x)$ of the regression function, independently of the parametric model assumptions; (ii) to compare the two estimates, $\hat{m}(x)$ and $m(x;\hat{\theta})$, by computing a suitable test statistic, followed by evaluation of its observed significance.

The first step has already been discussed in earlier chapters, both for the cases of continuous data and for the case of binary or count data through the idea of local likelihood. Comparison of the two fits, $\hat{m}(x)$ and $m(x;\hat{\theta})$, can be achieved by the likelihood ratio test statistic which represents the natural extension of an F statistic beyond the linear models context. This takes the form

$$\text{PLRT} = 2\sum_{i=1}^{n}\{\log f(y_i;\hat{m}(x_i),\hat{\psi}) - \log f(y_i; m(x_i;\hat{\theta}),\hat{\psi})\}.$$

In the case of a GLM, it is common to reformulate this statistic in the equivalent form

$$\text{PLRT} = D(y; m(x;\hat{\theta})) - D(y;\hat{m}(x)), \tag{5.3}$$

where $D(y;\mu)$ denotes the deviance function computed for the observed data y at the value of the mean μ. The possible presence of the dispersion parameter has not been made explicit in this notation.

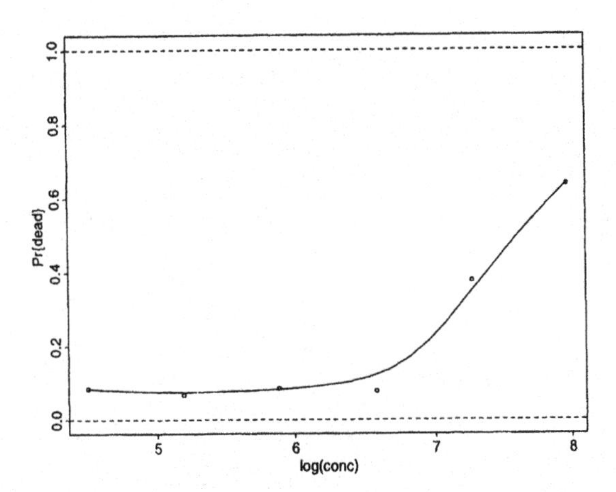

FIG. 5.7. The observed frequency of death in the trout data, plotted against log concentration of the toxicant, with a smooth regression curve.

For an illustration of this procedure consider Fig. 5.7, which refers to a dataset described by Hand *et al.* (1994, p. 340). This refers to an experiment on the effect of a toxicant on the survival of trout eggs, as observed 19 days after exposure. Only one of the two original treatment branches has been used for Fig. 5.7. The observed frequency of death of the trout eggs has been plotted against the concentration of the toxicant on a log scale. The smooth curve is the estimated probability of death, computed by the technique for smoothing binomial data described in Section 3.4.

For data of this sort, it is common to consider a logistic regression as a potential model. In the present case, this takes the form

$$\operatorname{logit}(\mathbb{P}\{\text{death}\}) = \alpha + \beta \log(\text{concentration}). \tag{5.4}$$

For binomial data, the deviance takes the form

$$D(y;\mu) = \sum_i y_i \log\left(\frac{p_i\,(1-\mu_i)}{(1-p_i)\,\mu_i}\right) + N_i \log\left(\frac{1-p_i}{1-\mu_i}\right),$$

where y_i, N_i, μ_i, p_i denote the number of events, the number of trials, the fitted mean value and the observed frequency y_i/N_i, respectively, at a given design point x_i.

The next step is to compute the significance of the observed value of the PLRT statistic. However, exact computations of the sort developed in Sections 5.2 and 5.3 are no longer feasible. A simple alternative is to simulate data from the fitted parametric model and compute nonparametric estimates $\hat{m}(x)$ from these simulated data, until a reasonably accurate estimate of the distribution of (5.3) under the fitted parametric model is obtained. The observed value of the statistic

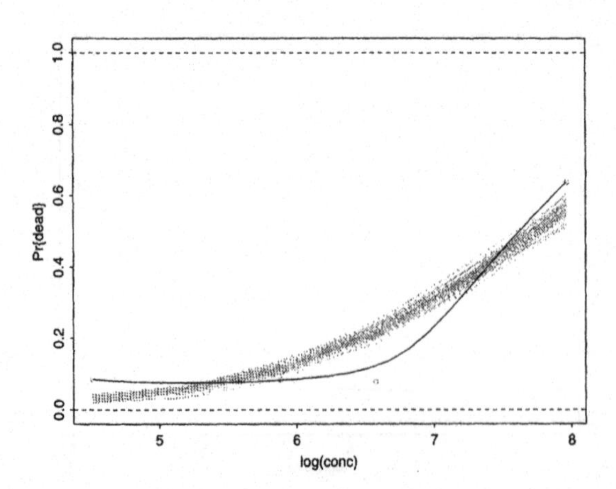

FIG. 5.8. A simulated reference band for a logistic model with the trout data. The visual impression of this figure is more effective on a computer screen than on paper.

can then be compared with this estimated null hypothesis distribution. This procedure is sometimes called the 'parametric bootstrap'.

This scheme is illustrated graphically in Fig. 5.8 which contains the same components of Fig. 5.7 with the addition of estimated regression curves produced from the simulated samples, represented by dashed lines. The superposition of these curves (50 of them in this case) produces the visual effect of a reference band similar to those obtained by more exact methods in earlier sections of this chapter. This graphical representation is not necessary for performing the pseudo-likelihood ratio test itself, but it is useful to illustrate its working, and for assessing where any departure from the model occurs.

The dashed lines of Fig. 5.8 are far away from the curve estimated from the original data, providing strong evidence against model (5.4). This conclusion is confirmed by the observed p-value of 0, out of 50 simulated samples.

A second illustration of the pseudo-likelihood ratio test in the context of GLMs is provided by the data reported by Bissell (1972) and plotted in Fig. 5.9. The data refer to the length (in metres) and the number of observed flaws in each of 32 rolls of cloth. Under simple assumptions, a plausible model is to regard the occurrence of flaws as generated by a Poisson process along the 'time' axis represented by the rolls. This leads to the parametric model

$$y \sim \text{Poisson}(\beta\, x), \tag{5.5}$$

where y denotes the number of flaws in a roll, x the length of the roll, and β an unknown positive parameter representing the intensity of occurrence of flaws.

This is in fact a GLM with a Poisson distribution for the response variable, a single covariate x, and an identity link function. This represents one of the very

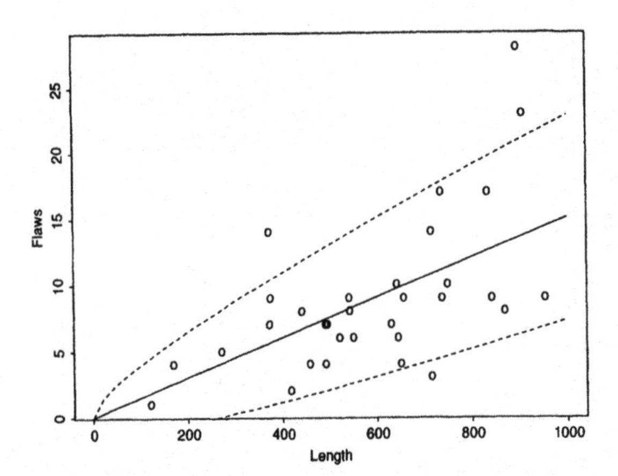

FIG. 5.9. A plot of the Bissell (1972) data with a parametric estimate of $m(x)$ and prediction intervals for the data.

few cases where a GLM allows explicit calculation of the maximum likelihood estimate, as

$$\hat{\beta} = \sum_i y_i / \sum_i x_i,$$

which defines the slope of the straight line displayed in Fig. 5.9.

An assessment of the appropriateness of the fitted model can be obtained by comparing the observed points with the expected variability of the data under the fitted model. This involves computing the so-called 'prediction intervals' defined by $\hat{\beta}x \pm 2\sqrt{\hat{\beta}x}$ in the present case. These are represented by dashed lines in Fig. 5.9. Notice that 4 out of 32 points fall outside these bands. Although this does not represent a formal test, it already raises some doubts about the validity of the proposed parametric model.

A nonparametric estimate of $m(x)$ can be constructed through the concept of local likelihood introduced in Section 3.4. In the present context, this takes the form

$$\ell_{[h,x]}(\beta) = \sum_i (y_i \log(\beta x_i) - \beta x_i) w_i,$$

where $w_i = w(x - x_i; h)$. Again, an explicit solution of the likelihood equation is feasible, namely

$$\hat{\beta}_x = \frac{\sum_i y_i w_i}{\sum x_i w_i},$$

with associated variance

$$\operatorname{var}\left\{\hat{\beta}_x\right\} = \beta \frac{\sum x_i w_i^2}{(\sum x_i w_i)^2}.$$

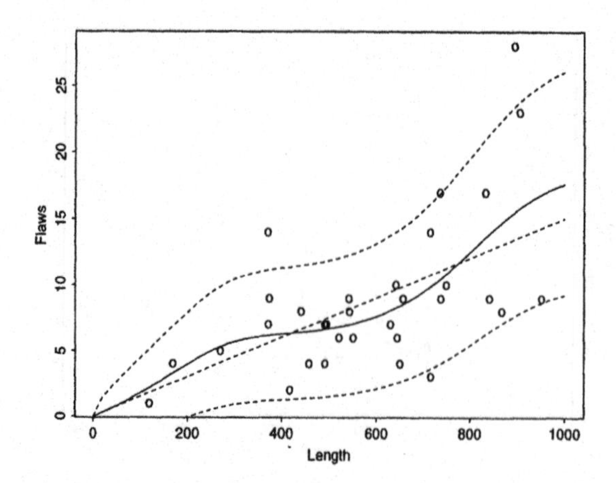

FIG. 5.10. The Bissell (1972) data with a nonparametric estimate of $m(x)$ and corresponding prediction intervals, for $h = 100$.

The continuous curve $\hat{m}(x) = \hat{\beta}_x x$ is created by repeating this computation for various values of x. Figure 5.10 also displays 'prediction intervals', for comparison with Fig. 5.9.

To assess the validity of the parametric model, the two estimates of $m(x)$, parametric and nonparametric, can be compared through the PLRT statistic. This leads to the difference of deviances

$$D(y; \hat{\beta}x) - D(y; \hat{m}(x)), \tag{5.6}$$

where

$$D(y; \mu) = \sum_i (\mu_i - y_i + y_i \log(y_i/\mu_i))$$

denotes the Poisson deviance for a given choice of μ.

Since the focus of interest is to check the parametric model $m(x) = \beta x$ and the Poisson assumption is not under consideration, Firth *et al.* (1991) suggested that this difference of deviances should be scaled by an estimate of the dispersion parameter ψ. Under the Poisson assumption this parameter is known to be 1. However, if this distributional assumption is not correct, the test statistic (5.3) would be affected, possibly showing significant departure from the model even if the assumed regression curve is correct. The scaled version

$$\frac{D(y; \hat{\beta}x) - D(y; \hat{m}(x))}{\hat{\psi}}, \tag{5.7}$$

where

$$\hat{\psi} = D(y; \hat{m}(x))/(n-1),$$

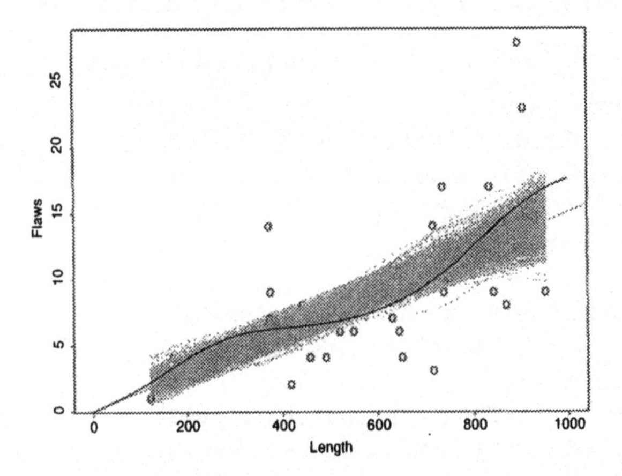

FIG. 5.11. A simulated reference band associated with the PLRT statistic for linearity, using $h = 100$, with the Bissell (1972) data.

has been proposed as a test statistic which is insensitive to failure of the Poisson assumption, and therefore focuses only on the assumed form of the regression function.

A graphical illustration of this modified statistic is provided by Fig. 5.11. For the observed data, the value of (5.7) is 5.30, when $h = 100$. An estimate of the significance, obtained via the simulation method described above, lies in the range 0.35 to 0.45, depending on the simulation run. These values do not vary much with h, in agreement with the pattern observed in other similar cases. This contrasts with use of the original form (5.3) which leads to a significance level around 0.06.

S-Plus Illustration 5.6. Trout data

Figure 5.8 was constructed with the following S-Plus *code.*

```
provide.data(trout)
conc <- N <- dead <-rep(0,6)
for(i in 1:6){
  conc[i] <- Concentr[i*4]
  for(j in 1:4){
      N[i] <- N[i] + Trouts[(i-1)*4+j]
      dead[i] <- dead[i] + Dead[(i-1)*4+j]
  }}
sm.logit.bootstrap(log(conc), dead, N, 0.5, nboot=50)
```

Figure 5.7 can be obtained by replacing the last line with

```
sm.logit(log(conc), dead, N, 0.5)
```

S-Plus Illustration 5.7. Bissell data: nonparametric estimate

Figure 5.10 was constructed with the following S-Plus *code.*

```
provide.data(bissell)
plot(Length, Flaws, xlim=c(0,1000), pch="o")
beta <- sum(Flaws)/sum(Length)
x <- seq(0, 1000, length=50)
lines(x, beta*x, lty=3)
h <- 100
W<-sm.weight(Length, x, h, poly.index=0)
sm.beta <- (W %*% Flaws)/(W %*% Length)
lines(x,sm.beta*x)
lines(x,sm.beta*x+2*sqrt(sm.beta*x),lty=3)
lines(x,sm.beta*x-2*sqrt(sm.beta*x),lty=3)
```

A script to produce Fig. 5.11 is also available. This is rather lengthy and has not been reproduced here. The main reason for the increased length of the script is that (5.5) implies a GLM with identity link function, while the functions `sm.poisson.bootstrap()` *and* `sm.logit.bootstrap()` *have been designed for testing parametric models with canonical links – logarithmic and logit respectively. Fortunately, the form of (5.5) allows explicit maximisation of the local likelihood, leading to relatively simple code. This script offers to the interested reader the possibility of examining in detail the working of the method.*

5.5 Further reading

This chapter is based on papers by Azzalini *et al.* (1989), where the PLRT criterion is presented, and Azzalini and Bowman (1993) where it is shown how to pursue exact numerical computation in the case of normal variates. Firth *et al.* (1991) proposed the modified form (5.7). Cleveland and Devlin (1988) give more general approximations for the distribution of F statistics, based on F distributions and approximate degrees of freedom. Weisberg and Welsh (1994) explored the form of the link function in generalised linear models. Related work has also been done by Cook and Weisberg (1994).

Cox *et al.* (1988), Cox and Koh (1989) and Eubank and Spiegelman (1990) developed methods for the problems described in this chapter, based on spline and orthogonal series estimators.

Le Cessie and van Houwelingen (1991; 1993; 1995) discuss a variety of issues associated with the use of nonparametric methods to assess the fit of a regression model, with particular emphasis on logistic regression.

Landwehr *et al.* (1984), Fowlkes (1987) and Staniswalis and Severini (1991) discuss graphical diagnostics, based on smoothing, for regression models.

Diblasi and Bowman (1997) show how a test of no effect can be applied to residuals from a linear model to check the assumption of constant variance.

Exercises

5.1 *Great Barrier Reef data.* The S-Plus illustrations of Sections 5.2 and 5.3 have considered subsets of the Great Barrier Reef data. By selecting other suitable values of the `Zone` and `Year` variables, try out the methods on other subsets of this dataset.

5.2 *Radiocarbon dating.* Reconsider the data of Fig. 4.2, for which a preliminary assessment of linearity has been obtained via the use of variability bands. Produce a significance trace of the hypothesis of linearity for these data.

5.3 *Your data.* Regression with normal errors is one of the most commonly applied techniques. Recall data you have encountered, involving y and a single x, where you were not entirely satisfied with the linearity assumption. Try the methods of Section 5.3 on these data. Consider in particular the relevance of the choice of the bandwidth on the p-value of the test statistic, and more generally on the overall conclusions. Do the same for data with two explanatory variables, of the form (x_1, x_2, y). Examine whether the significance trace provides consistent indication over a reasonable range of h values.

5.4 *Discussion.* The method presented in Section 5.3 is nonparametric in nature, but it assumes normality of the errors ε. Discuss whether having a nonparametric method based on such a specific parametric assumption is a self-defeating approach. Consider aspects such as robustness of the F ratio to departure from normality (in various forms), scientific relevance of various modes of departures from the 'ideal' form of standard linear regression, and possible extensions of the distribution theory presented here to more general settings.

5.5 *Exact computation.* In Section 5.3, an approximate computation of the p-value is presented, and this is adopted as the standard method in the `sm` library. If you do not like approximations, try exact computation using the distribution theory mentioned in the text, and implementing it with the aid of your favourite programming language. Examine the difference between the two computation methods on some dataset with moderate sample size. (This saves computing time and it also makes differences generally more prominent.) Are these differences relevant for statistical analysis?

5.6 *Quadratic logistic regression.* The linear logistic model (5.4) is clearly inadequate for the trout data. An extended model with a quadratic component for the explanatory variable log(concentration) is the next natural candidate. Examine the effect of this by adding the parameter `degree=2` to the function `sm.logit.bootstrap()` in the script listed in S-Plus Illustration 5.6.

5.7 *Variability bands.* Figure 5.10 displays prediction intervals for $m(x)$ from Bissell's data. Construct a similar plot with variability bands, similarly to Figure 4.2.

5.8 *PLRT via simulations.* Consider some values of h in a reasonable range and, for each of them, execute the S-Plus code which performs the PLRT to assess linearity in Bissell's data. Plot the corresponding significance trace. Does this curve provide a consistent view of the evidence on linearity?

5.9 *Modification of the PLRT.* The appropriateness of the scaled version (5.7) of (5.3) has not been fully explored in the literature. Examine the effect of this modification in other cases. In particular, write a modified form of `sm.logit.bootstrap()` which implements the modification (5.7), apply this to the trout data similarly to S-Plus Illustration 5.6, and comment on the results. If you are very brave, attempt a general solution to the scaling problem.

6

COMPARING CURVES AND SURFACES

6.1 Introduction

In Chapter 4 some simple inference was carried out by assessing whether a covariate had any effect on a response. In Chapter 5 the special case of comparing nonparametric and parametric models was discussed. This topic was also mentioned in Chapter 2 where a normal shape for densities was considered. In this chapter nonparametric models for different groups of data are compared. Both density estimates and nonparametric regression curves are considered.

6.2 Comparing density estimates in one dimension

The aircraft data were discussed in some detail in Chapters 1 and 2. The motivation for collecting these data was largely to provide a description of the way in which aircraft technology has developed during the twentieth century. The analysis so far has therefore been largely in terms of graphics, providing some insight into the directions in which development has taken place. Comparisons were drawn among three major groups, corresponding to the time periods 1914–1935, 1936–1955 and 1956–1984.

Figure 6.1 compares the density estimates for the log of wing span across these three groups, making the comparisons in pairs for clarity, and to focus on the evolution over time. Each of the estimates has a shoulder on the right hand side of the density, although these features are of varying sizes. It would be helpful to have a means of identifying whether the differences between the estimates reflect systematic differences in the underlying distributions, or whether they could be attributed simply to random variation. With these data, the aims are descriptive and independent random sampling is not a very plausible model. Standard errors and formal significance tests are therefore of limited value, although this approach may be of some use in providing a natural scale against which comparisons can be plotted.

In a formal approach, the hypotheses are

$$\begin{aligned} H_0 &: f(y) = g(y), \quad \text{for all } y, \\ H_1 &: f(y) \neq g(y), \quad \text{for some } y. \end{aligned}$$

In Section 2.5, an integrated squared error statistic was proposed for comparing a density estimate with a normal curve. An analogous approach for comparing two density estimates $\hat{f}$ and $\hat{g}$ is through the statistic

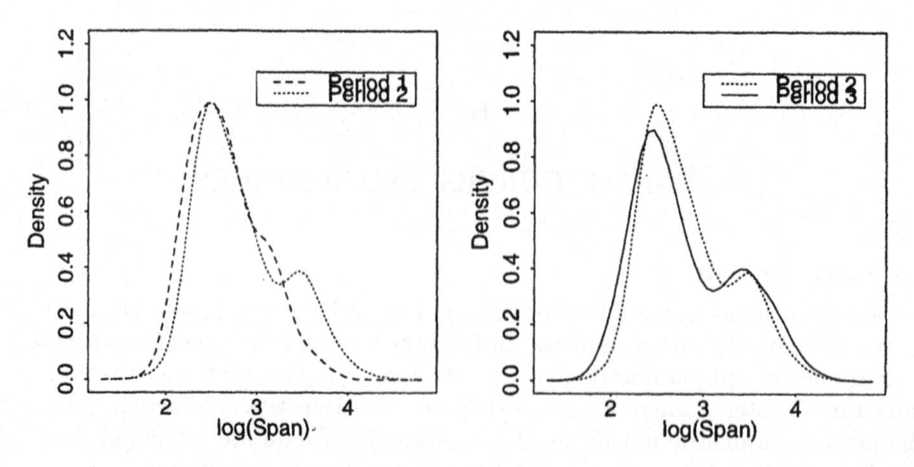

FIG. 6.1. Density estimates for the aircraft span data, on a log scale. The left panel compares groups 1 and 2; the right panel compares groups 2 and 3.

$$\int \{\hat{f}(y) - \hat{g}(y)\}^2 dy. \tag{6.1}$$

In the case of a test of normality, when normal kernel functions are used, it was shown how the integration can be carried out analytically, since the statistic can be represented as the sum of convolutions of normal curves. The same strategy can be applied to expression (6.1). However, when sample sizes are large it is more efficient, and sufficiently accurate, to approximate the value of the integral through numerical integration, using density estimates which are evaluated over a finely spaced grid along the horizontal axis.

When estimates of the densities are constructed individually this will lead to the use of two different smoothing parameters. There is, however, some potential advantage to be gained in using the same smoothing parameter for each estimate. This arises from the properties of density estimates, which were briefly described in Section 2.2. In particular, the means of the estimates are

$$\mathbb{E}\left\{\hat{f}(y)\right\} = \int \phi(y - z; h)\, f(z)\, dz,$$

$$\mathbb{E}\{\hat{g}(y)\} = \int \phi(y - z; h)\, g(z)\, dz.$$

Under the null hypothesis that the two density functions f and g are identical, these two means will therefore also be identical if the same smoothing parameter h is used in the construction of each. The contrast of the estimates $\hat{f} - \hat{g}$ will have mean 0 under the null hypothesis. For the three groups of aircraft span data, the normal optimal smoothing parameters are 0.14, 0.16 and 0.19 respectively. The geometric mean of these, namely 0.16, therefore offers a suitable compromise which ensures this attractive property in the comparison of the estimates.

The distributional properties of the test statistic are difficult to establish when the null hypothesis is of such a broad form, with no particular shape specified for the common underlying density. A similar situation holds with a more familiar form of nonparametric testing, when ranks are used as the basis of analysis. In a two-sample problem the distribution of the sum of the ranks of one set of observations within the joint sample can be calculated easily. A similar approach in the present setting is to condition on the observed data and to permute the labels which identify each observation with a particular group. Under the null hypothesis, the allocation of group labels is entirely random, since both groups of data are generated from the same underlying density function. The distribution of statistic (6.1) under the null hypothesis can therefore be constructed easily by evaluating the statistic on data produced by random permutation of the group labels. The empirical significance of the statistic calculated from the original data is then simply the proportion of simulated values which lie above it.

When this process is applied to the aircraft span data, the empirical p-value is 0, with a simulation size of 100. This therefore offers clear evidence that there are systematic differences between the densities for groups 1 and 2. If this process is repeated for groups 2 and 3, an empirical p-value of 0.02 is obtained, suggesting again that significant differences are present.

Where more than two groups are to be compared, a suitable statistic is

$$\sum_{i=1}^{p} n_i \int \{\hat{f}_i(y) - \hat{f}(y)\}^2 dy,$$

where $\hat{f}_1, \ldots, \hat{f}_p$ denote the density estimates for the groups, $\hat{f}$ denotes the density estimate constructed from the entire set of data, ignoring the group labels, and n_i denotes the sample size for group i. In order to preserve zero bias in the comparisons, a common smoothing parameter should again be employed, including in the combined estimate $\hat{f}$.

In interpreting the fact that the density estimates for groups 1 and 2 are significantly different, there are at least two principal features which could be the cause of this. One is the stronger right hand shoulder in group 2 and the other is the higher density for group 1 in the left hand tail. It is helpful to follow up the global significance test with a graphical display of the type used in Chapter 5, where nonparametric and parametric curves are compared. In the present setting, a reference band should allow two density estimates $\hat{f}$ and $\hat{g}$ to be compared. At any point y, the quantity $\text{var}\left\{\hat{f} - \hat{g}\right\}$ is therefore crucial. In Section 2.3, the advantages of working on the square root density scale were described, since this corresponds to the variance stabilising transformation. This helpful feature can be employed in the present setting by comparing $\sqrt{\hat{f}}$ and

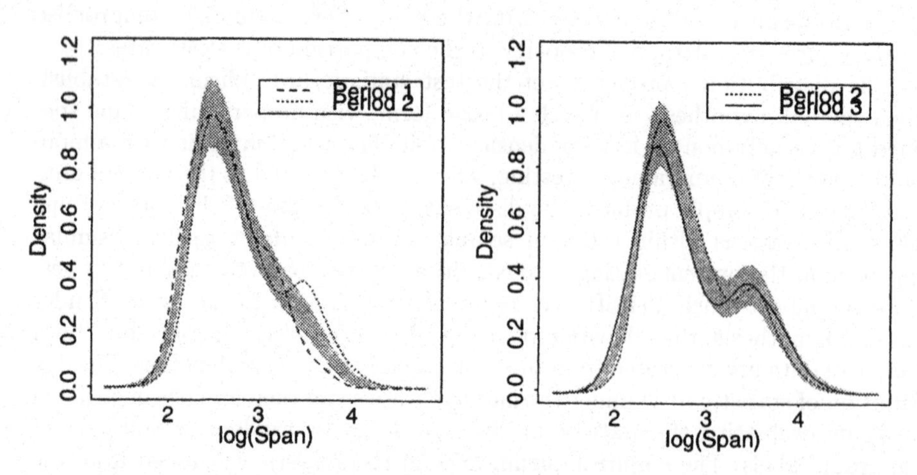

FIG. 6.2. Density estimates for the aircraft span data, on a log scale, with a reference band for equality. The left panel compares groups 1 and 2; the right panel compares groups 2 and 3.

$\sqrt{\hat{g}}$, for which the relevant variance is

$$\begin{aligned} \text{var}\left\{\sqrt{\hat{f}} - \sqrt{\hat{g}}\right\} &= \text{var}\left\{\sqrt{\hat{f}}\right\} + \text{var}\left\{\sqrt{\hat{g}}\right\} \\ &\approx \frac{1}{4}\frac{1}{nh}\int w^2(y)dy + \frac{1}{4}\frac{1}{nh}\int w^2(y)dy \\ &= \frac{1}{2}\frac{1}{nh}\int w^2(y)dy, \end{aligned}$$

with the square root of this expression giving the standard error.

At any point y, interest would be raised when the square root densities differed by more than two standard errors. A simple means of displaying the separation graphically is to superimpose on a plot of the estimates a reference band which is centred at the average of the two curves and whose width at the point y is two standard errors. This can easily be re-expressed on the original density scale. Reference bands of this type for the aircraft span data are displayed in Fig. 6.2. For groups 1 and 2 it is apparent that there are major differences in both the left hand and right hand tails of the distributions. For groups 2 and 3 it is confirmed that it is the left hand tail of the density function which is the major source of difference between the two distributions.

When distributions are compared through rank tests it is often assumed that each has the same shape and that differences lie only in location. This structure can also be examined through density based tests. In fact, a test of shape can be standardised for scale as well as location. Figure 6.3 displays density estimates constructed from the aircraft span data, where each group has first been standardised by sample mean and standard deviation. The permutation test

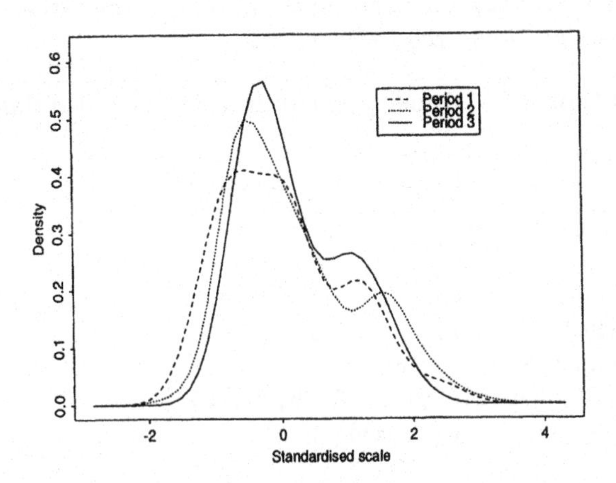

FIG. 6.3. Density estimates for the aircraft span data, standardised for location and scale.

can now be applied to the standardised data. This leads to an empirical p-value of 0, with a simulation size of 100, confirming that the differences among the groups are more than simple location and scale changes from some common shape. Strictly speaking, the process of estimation implicit in the use of the sample mean and standard deviation should also be incorporated into the permutation distribution. However, this effect is likely to be small, particularly where the sample sizes are large, as in this case. Since the procedure is a computational one, other forms of standardisation, such as medians or modes, and robust estimates of scale parameters, can easily be used.

S-Plus Illustration 6.1. Density estimates for the log span data

The following S-Plus *code may be used to reconstruct Fig. 6.1.*

```
provide.data(aircraft)
par(mfrow = c(1,2))
h <-  exp(mean(log(tapply(log(Span), Period, FUN = "hnorm"))))
ind <- (Period!=3)
sm.density.compare(log(Span)[ind], Period[ind], h = h,
        xlab = "log(Span)", lty = c(3,2), ylim = c(0, 1.2))
legend(3.0, 1.1, c("Period 1", "Period 2"), lty = c(3,2))
ind <- (Period!=1)
sm.density.compare(log(Span)[ind], Period[ind], h = h,
        xlab = "log(Span)", lty = c(2,1), ylim = c(0, 1.2))
legend(3.0, 1.1, c("Period 2", "Period 3"), lty = c(2,1))
par(mfrow = c(1,1))
```

Figure 6.2 can be reproduced by adding the arguments `model="equal"` *and* `test=F` *to the* `sm.density.compare` *function. The permutation tests can be carried out simply by adding* `model="equal"`.

S-Plus Illustration 6.2. Density estimates for the standardised span data

---

The following S-Plus *code may be used to reconstruct Fig. 6.3.*

```
provide.data(aircraft)
y <- log(Span)
for (i in 1:3) {
  yi  <- y[Period == i]
  med <- median(yi)
  sc  <- diff(quantile(yi, c(0.25, 0.75))) / 1.349
  y[Period == i] <- (yi - med) / sc
  }
h <-  exp(mean(log(tapply(y, Period, FUN = "hnorm"))))
sm.density.compare(y, Period, h = h, lty = c(3,2,1),
        xlab = "Standardised scale")
legend(1.5, 0.55, c("Period 1","Period 2","Period 3"),
        lty=c(3,2,1))
```

6.3 Relative risk in two dimensions

The ideas described in the previous section all extend in principle to higher dimensions. A particular application arises in two dimensions, in epidemiological problems where the spatial positions of disease occurrence are recorded and where interest lies in identifying whether the risk of contracting the disease changes over the geographical region. Figure 6.4 displays an example concerning the occurrence of laryngeal cancer, which has been analysed by a number of different investigators. Data, adjusted to preserve anonymity, are provided by Bailey and Gatrell (1995). The left panel of the figure displays the spatial positions of the cases which have occurred throughout the period 1974–1983 in a region of the north-west of England. The possible effect of an incinerator, now disused, and marked on the figure by a triangle, in increasing the risk of laryngeal cancer was of particular interest. The pattern of occurrence of cases will clearly be closely related to the general spatial pattern of the population in that region. In order to provide a control pattern, data on lung cancer cases were also collected. These data should also exhibit structure determined by the population patterns, and the use of lung cancer patients is intended to define a population which is likely to have similar age, sex and other characteristics to the laryngeal cancer population. The problem of identifying spatial variation in risk then reduces to comparing the case and control patterns of occurrence.

Bithell (1990) described a similar set of spatial data on childhood leukaemia and proposed the use of two-dimensional density estimates to describe and com-

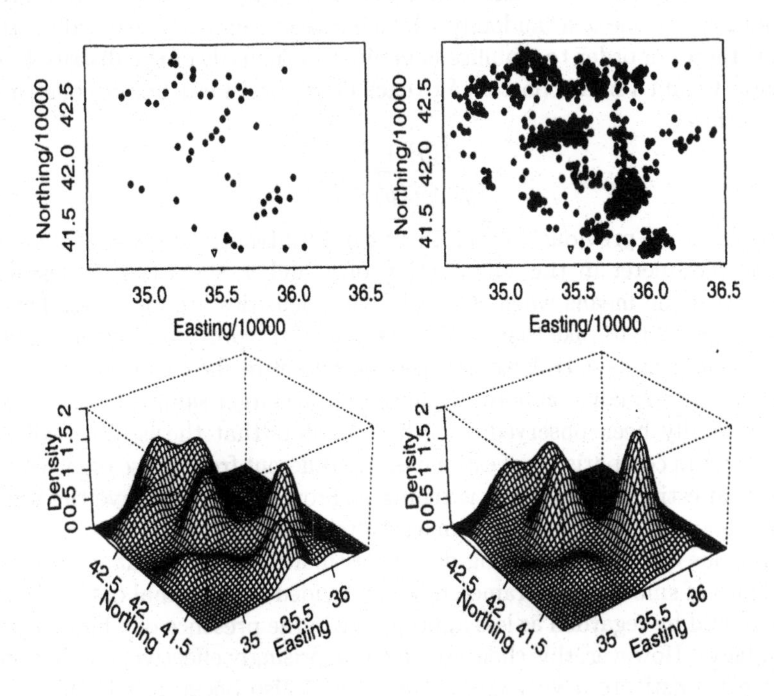

FIG. 6.4. Spatial positions and density estimates for the laryngeal cancer data. The left panels show the cases and the right panels the controls. The position of the incinerator is marked by a triangle.

pare the patterns exhibited by the data. It would be natural to think in terms of a point process with some unspecified intensity function as a model for problems of this type. However, this reduces to a problem involving density functions by conditioning on the total number of data points observed in each of the case and control groups. Figure 6.4 displays the case and control density estimates for the laryngeal cancer data. While some broad features of the data are apparent, the high degree of multimodality in the distributions, and the three-dimensional perspectives of the figures, make detailed graphical comparison in this form rather difficult.

Bithell (1990) proposed that a relative risk function should be constructed as $\rho(y) = f(y)/g(y)$, where f denotes the density of the cases, g the density of the controls, and y is a two-dimensional quantity indicating spatial position. He also proposed the convenient transformation

$$\frac{\rho(y)}{1+\rho(y)} = \frac{f(y)}{f(y)+g(y)},$$

which is the conditional probability of a case appearing in a small region near y, given that exactly one of a randomly selected case or randomly selected control is to appear there. In order to stabilise behaviour in the tails of the distributions it is advisable to add a small constant δ to each density estimate before constructing the risk surface:

$$\frac{\hat{f}(y)+\delta}{\hat{f}(y)+\hat{g}(y)+2\delta}.$$

As discussed in the previous section, there can be advantages in using a common smoothing parameter in the construction of $\hat{f}$ and $\hat{g}$, since the two estimates will have the same means when the underlying densities are identical. This is a rather stronger step to take than in the example of the previous section, since the numbers of cases and controls are markedly different in the cancer data, and it is usual to employ a smaller smoothing parameter for larger sample sizes. However, it has repeatedly been observed in earlier chapters that the issues involved in inference, and in comparisons of estimates, are different from those involved when a single good estimate of a curve is required. From this perspective the use of a common smoothing parameter has much to recommend it.

In previous examples involving density estimation some guidance on appropriate values of smoothing parameters was provided by a normal optimal choice. Here this could be regarded as less appropriate in the presence of a high degree of multimodality. However, this choice remains surprisingly effective, partly because the controls constitute a very large sample, and also because it is the ratio of densities, rather than the densities themselves, which are of interest. In addition, the geographical nature of the sample space makes it appropriate to use the same smoothing parameter in each co-ordinate direction. This leads to the suggested values of $(0.12, 0.12)$ for (h_1, h_2), which is comparable with the smoothing parameters used by other authors; see the discussion in Anderson and Titterington (1997).

Figure 6.5 displays an estimated relative risk surface for the laryngeal cancer data, using $\delta = 0.1$. In order to symmetrise the scale, a log transformation has also been used. Both surfaces show considerable variation although this does not appear to have any simple, systematic structure. There is also no natural scale of reference against which to compare the size of the observed variation. An alternative approach is to construct the simple difference of the density estimates, as used in the previous section. This is a much less natural than the ratio, or log ratio, scales but it has the advantage that the square root variance stabilising transformation allows a simple assessment of variability. In particular, the difference between the square root density estimates at any point can be measured in terms of standard deviations. A simple display of a reference band is difficult to construct for a three-dimensional plot. It is therefore informative to plot the surface

$$\frac{\sqrt{\hat{f}(y)}-\sqrt{\hat{g}(y)}}{\sqrt{\alpha(w)^2/(4nh^2)}},$$

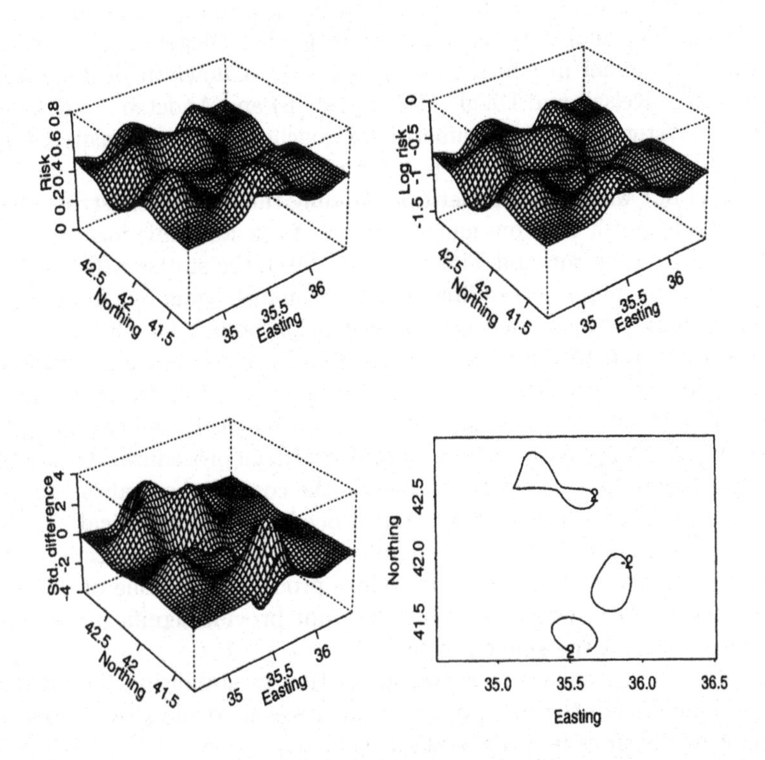

FIG. 6.5. Comparison of case and control patterns for the laryngeal cancer data. The top left panel shows the risk surface while the top right panel shows this on a log scale. The bottom panels display the standardised density difference surface.

where the expression under the square root sign in the denominator is the bivariate version of the variance derived in the previous section, using the multivariate approximation stated in Section 2.2. This surface is displayed in the bottom two panels of Fig. 6.5. Following the usual argument for a reference band, which indicates how far apart the two densities will generally be when the null hypothesis of equality is true, attention can now be focused on those positions y where the standardised difference between the density estimates is markedly larger than 2, in absolute value. There are two positions of particular interest where the density difference exceeds two standard deviations. The position at the foot of the plot, where the excess of case over control density is more than two standard deviations, lies close to the location of the incinerator, and is clearly due to the small cluster of cases identified by Diggle (1990). However, it is notable that there is also a difference of more than two standard deviations in the opposite direction at a nearby location.

Techniques of this type are essentially graphical and exploratory. A more formal examination of the evidence for spatial variation in risk can be conducted by fitting and analysing an appropriate model. Diggle (1990) proposed a semiparametric model in which risk decreases as a function of distance from the incinerator. Kelsall and Diggle (1995a; 1995b) and Anderson and Titterington (1997) investigated a nonparametric approaches based on density estimates. These approaches will have lower power than any test based on a valid parametric model but will have the merit of avoiding the need to construct the form of such a model, with the consequent danger of selecting this form inappropriately. Following Anderson and Titterington (1997), the statistic (6.1) can easily be constructed to compare two-dimensional estimates, again by numerical integration using a two-dimensional grid of evaluation points. On this occasion the relationship between the case and control distribution is not a symmetric one, and it is natural to regard the control density as representing the null hypothesis. New case data can therefore be simulated from this density, under the null hypothesis, in what is referred to as a *smoothed bootstrap* procedure. An alternative approach, adopted here, is simply to sample the control observations to provide new case data, in an ordinary bootstrap procedure. The empirical significance of the observed data can then be computed as in the previous section. With a bootstrap sample size of 100, this procedure produces a p-value of 0.71 for the laryngeal cancer data, which therefore does not provide significant evidence of spatial variation over the whole region.

The effect of the square root transformation in stabilising the variance of density estimates leads to the suggestion that this scale would also be appropriate for comparing densities in a test statistic such as

$$\int \left\{ \sqrt{\hat{f}(y)} - \sqrt{\hat{g}(y)} \right\}^2 dy.$$

The reader is invited to explore the effect of this in an exercise at the end of the chapter.

Kelsall and Diggle (1995a; 1995b) also discuss a nonparametric approach to the modelling of spatial variation in this dataset. In particular, they derive an interesting and useful procedure for selecting appropriate smoothing parameters when interest lies in the log ratio of density estimates and they discuss the potential merits of using a common smoothing parameter. Testing of the hypothesis of constant risk, and the construction of an appropriate reference band are also carried out on the laryngeal data, using the distance of each observation from the incinerator as a one-dimensional position variable.

The highly multimodal nature of the underlying density functions suggests that a variable bandwidth approach might be beneficial with these data. However, the derivation of standard errors for the difference between the estimates will be more complex.

S-Plus Illustration 6.3. Plots and density estimates from the laryngeal cancer data

The following S-Plus *code may be used to reconstruct Fig. 6.4.*

```
provide.data(lcancer)
cases    <- cbind(Easting, Northing)[Cancer == 1,]/10000
controls <- cbind(Easting, Northing)[Cancer == 2,]/10000
xlim     <- range(Easting/10000)
ylim     <- range(Northing/10000)
par(mfrow=c(2,2))
plot(Easting/10000, Northing/10000, type = "n")
points(cases)
points(35.45, 41.3, pch = 6)
plot(Easting/10000, Northing/10000, type = "n")
points(controls)
points(35.45, 41.3, pch = 6)
h <- c(0.12,0.12)
sm.density(cases,    h = h, xlim=xlim, ylim=ylim, zlim=c(0,2))
sm.density(controls, h = h, xlim=xlim, ylim=ylim, zlim=c(0,2))
par(mfrow=c(1,1))
```

A longer script to reconstruct Fig. 6.5 is also available.

6.4 Comparing regression curves and surfaces

Fishing data from the Great Barrier Reef were introduced in Section 3.3, where nonparametric curves were used to explore the data and to identify that longitude, broadly associated with distance from the shore, is an important covariate. One of the aims in collecting these data was to identify whether there are any differences between the closed and open zones in the abundance of marine life found near the sea bed. A suitable model is therefore

$$y_{ij} = m_i(x_{ij}) + \varepsilon_{ij},$$

where i is a zone indicator and j indexes the n_i observations within each zone. Interest lies in identifying whether the regression functions for the two zones, m_1 and m_2, are different. Figure 6.6 displays the data and regression curves for the 1992 survey. The catch score broadly decreases with longitude, with the closed zone producing a consistently higher score, by a small margin, at larger values of longitude. Some caution has to be exercised again on the assumption of normality where some observations have reached the limits of detection near the end of the longitude scale.

Hall and Hart (1990) considered data of this type where the observations were made at the same covariate values for each group. This simplifies the problem by allowing differences between the responses to be computed. King *et al.* (1995) compared the estimated regression curves and used a simulation test to assess

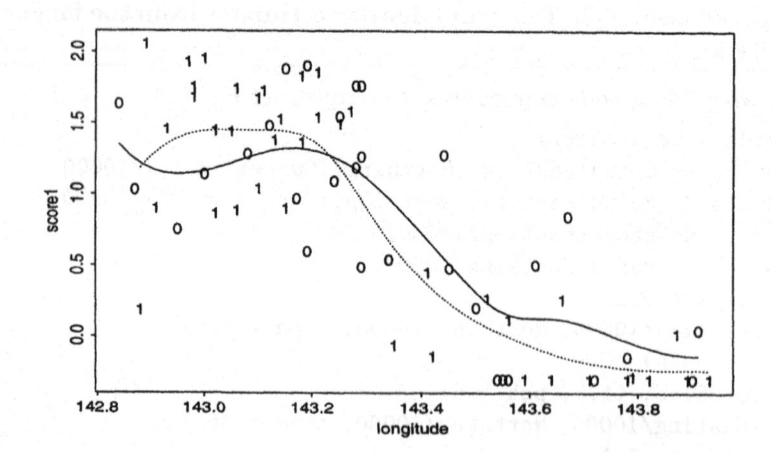

FIG. 6.6. Regression curves for the open (dotted line) and closed (full line) zones with the Great Barrier Reef data.

differences. Comparisons of this type are particularly effective with local linear regression estimates as a result of the form of the bias given in (4.4). Bias at any point x is controlled principally by the size of the smoothing parameter and the shape of the true curve rather than the particular pattern of design points in that neighbourhood, and so comparisons between the regression estimates are simplified. In particular, when the underlying curves are identical, or even simply have the same shape, then the biases will be identical if a common smoothing parameter is used.

A suitable statistic for comparing regression curves is

$$\frac{\sum_{i=1}^{p}\sum_{j=1}^{n_i}\{\hat{m}_i(x_{ij}) - \hat{m}(x_{ij})\}^2}{\hat{\sigma}^2}, \tag{6.2}$$

where $\hat{m}_i$ denotes the regression estimate from group i, $\hat{m}$ denotes a regression estimate constructed from the entire dataset, assuming that all groups have identical regression functions, and $\hat{\sigma}^2$ denotes an estimate of the error variance σ^2. This form of statistic provides a means of comparing any number of regression curves.

The distribution of this test statistic can be computed using the results on quadratic forms which were described in Chapters 4 and 5. In matrix notation, the vector of fitted values $\hat{m}_i$ from each individual regression estimate can be written as $\hat{m}_i = S_i y_i$, where S is an $n_i \times n_i$ matrix of weights and y_i denotes the vector of responses for the ith group. The entire collection of fitted values can therefore be gathered together in the form $\hat{m} = S_d y$, where the subscript d indicates that smoothing is being performed under the model which assumes the regression curves to be different. Similarly, the fitted values under the model which assumes the regression curves to be the same can be computed by pooling the data and represented as $\hat{m} = S_s y$, for a matrix of known weights S_s. The numerator of the test statistic (6.2) is therefore $y^\top (S_d - S_s)^\top (S_d - S_s) y$, which is a quadratic form.

A simple estimator of σ^2 is provided by the differencing approach described in Section 4.3. This can also be expressed as $y^\top By$ for a matrix of constants B. This is therefore also a quadratic form in the data, and so the test statistic can be written as

$$\frac{y^\top Qy}{y^\top By}.$$

As discussed in Section 4.3, the effect of bias is to inflate the estimate of σ^2. This leads to underestimation of the difference between the curves, and a conservative test procedure.

Calculation of the p-value associated with the test therefore proceeds by the techniques described in Sections 5.2 and 5.3. The quadratic form calculation is equivalent to one based on ε rather than y because, due to the bias properties of the local linear estimator, the means of $\hat{m}_i(x)$ and $\hat{m}(x)$ are approximately equal. For the estimates displayed in Fig. 6.6 the p-value is 0.57, using the smoothing parameter $h = 0.1$. The p-value remains well above 0.05 for all values of h. There is therefore no real evidence of a difference between the patterns of catch score.

Reference bands can be constructed to follow up the results of the significance tests. The standard error associated with a regression estimate was discussed in Chapter 4, and so the standard error of the difference of estimates can easily be computed as

$$se = se\{\hat{m}_1(x) - \hat{m}_2(x)\} = \sqrt{se_1(x)^2 + se_2(x)^2},$$

where $se_1(x)$ and $se_2(x)$ denote the standard errors of the individual curves. Under the assumption that the regression curves are identical, the means of the two estimates are also identical, and so the difference between the two estimates has mean 0. At any point x, interest will be raised where the estimates are more than two standard errors apart. A reference band of width two standard errors, centred at the average of the two curve estimates, can therefore be superimposed on a plot to aid the comparison of the curves. This has been done in Fig. 6.7, which confirms the absence of evidence of differences between the open and closed zones.

This technique for comparing regression curves can be extended to two dimensions to compare regression surfaces constructed from two covariates. The numerator of statistic (6.2) produces a quadratic form, since the estimates continue to have the structure $S_i y_i$, for a suitable matrix of weights S_i, as discussed in Section 4.2. The simple difference based estimator of σ^2 does not immediately extend to two dimensions, but the residual sum of squares from a regression surface constructed with small values of the smoothing parameters provides an effective alternative.

As an illustration of the two-dimensional case, Fig. 6.8 displays the reef data using both latitude and longitude as covariates. The top two panels of the figure

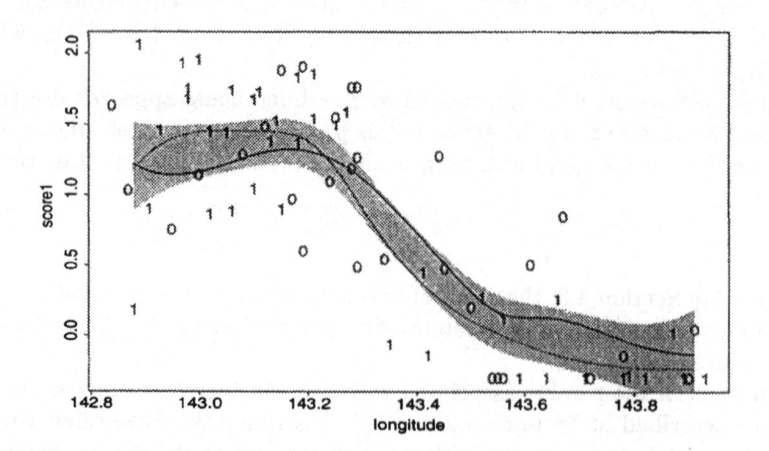

FIG. 6.7. Reference bands for equality between the closed and open zones with the Great Barrier Reef data for the 1993 survey.

refer to surveys carried out in 1992 (left) and 1993 (right). Some small differences are apparent, but it is difficult to make a visual assessment of whether these indicate systematic differences or random variation. Judgements of this type are particularly difficult to make when the data are not indicated on the plot. Since the estimates of the two regression surfaces are linear combinations of the response variables, and since the local linear approach in two dimensions retains the helpful property that bias does not depend on the pattern of the design points, the techniques of formal testing for equality, and of constructing a reference band, extend from the one-dimensional case.

A formal test will be discussed in Chapter 8. Here a graphical comparison of the surfaces is constructed to provide informal guidance. The lower panels of Fig. 6.8 display the surface

$$\frac{\{\hat{m}_1(x) - \hat{m}_2(x)\}}{se\{\hat{m}_1(x) - \hat{m}_2(x)\}}.$$

The standard error in the denominator is computed in the same way as for the one-dimensional case, with an estimate of the error variance σ^2 based on a residual sum of squares and approximate degrees of freedom, as discussed in Chapter 4. The resulting surface of standardised differences lies mostly within the range ± 2. However, there are two regions where the difference between the estimates is more than two standard deviations, indicating possible structure which was not apparent from the plots of the original surfaces.

The techniques described in this section also extend to binary and binomial response data. Weidong *et al.* (1996) report a survey to assess the prevalence of a human parasitic worm infection in a rural village in China. Figure 6.9 displays nonparametric estimates of the proportions infected as a function of age,

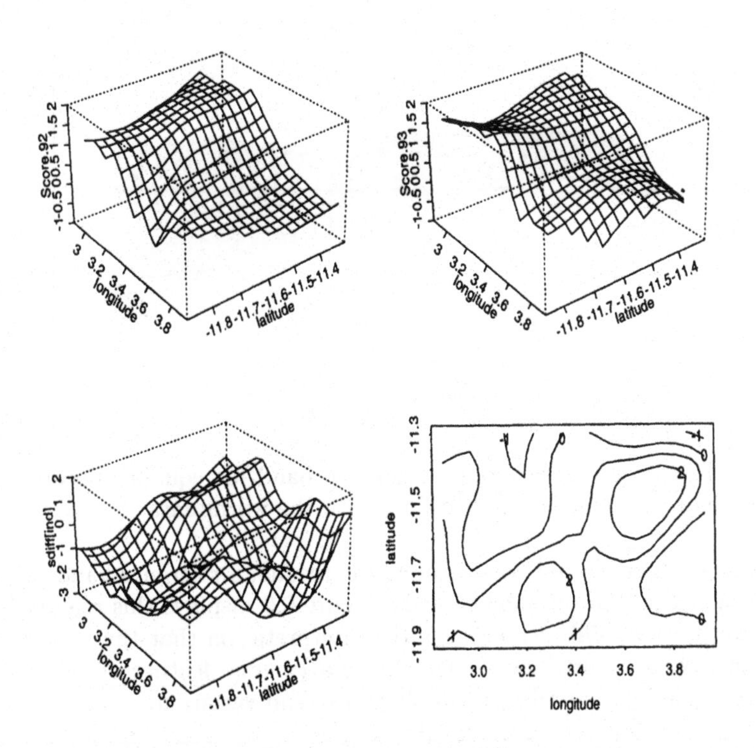

FIG. 6.8. Regression surfaces for 1992 and 1993 (top panels) and the standardised difference (bottom panels) for the Great Barrier Reef data. In order to achieve more attractive scales, 143 has been subtracted from longitude.

separately for males and females. These curves were produced by the local likelihood technique discussed in Section 3.4. The comparison suggests that children of both sexes have similar levels of infection, but that the infection rate for females does not decrease with age as rapidly as it does for males. Since the data become increasingly sparse with age, this is an interpretation which needs to be backed up by firmer evidence than visual assessment. Young (1996) used quadratic forms to derive a formal test of equality for regression curves with binary data, but this is more awkward to implement than in the normal errors case. A permutation or bootstrap test, where group membership is randomly permuted across the infection–age pairs, is straightforward in principle but computationally rather intensive. For the worm data, where the aims are essentially descriptive, a reference band will suffice for present purposes.

The standard errors associated with a local logistic regression curve were discussed in Section 4.4. These are most naturally expressed on the logit scale. By combining these on the logit scale, following the steps described above for

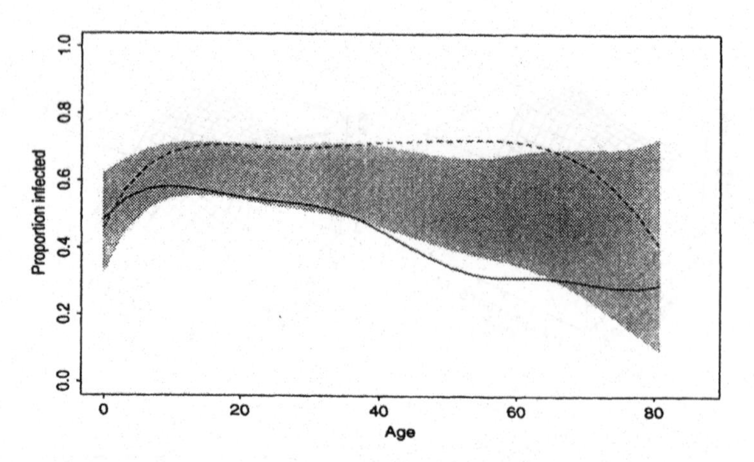

FIG. 6.9. Regression curves and a reference band for equality between males (solid line) and females (dashed line) with the worm data.

the normal errors case and transforming the results, a reference band can easily be constructed on the probability scale. Figure 6.9 displays this reference band for the worm data. This strengthens the interpretation that the slower rate of decline in infection with age for females is a genuine feature of the data and cannot reasonably be attributed simply to random variation.

S-Plus Illustration 6.4. Regression curves for the reef data

The following S-Plus code may be used to reconstruct Fig. 6.6.

```
provide.data(trawl)
ind       <- (Year == 0)
longitude <- Longitude[ind]
zone      <- Zone[ind]
score1    <- Score1[ind]
sm.ancova(longitude, score1, zone, h = 0.1)
```

S-Plus Illustration 6.5. Reference bands for equality in the reef data

The following S-Plus code may be used to reconstruct Fig. 6.7.

```
provide.data(trawl)
ind       <- (Year == 0)
longitude <- Longitude[ind]
zone      <- Zone[ind]
score1    <- Score1[ind]
sm.ancova(longitude, score1, zone, h = 0.1, model = "equal")
```

S-Plus Illustration 6.6. Regression curves and a reference band for equality in the worm data

The following S-Plus *code may be used to reconstruct Fig. 6.9.*

```
provide.data(worm)
Males    <- sm.logit(Age[Sex == 1], Infection[Sex == 1], h = 10,
               display = "none")
age     <- Males$eval.points
Females <- sm.logit(Age[Sex == 2], Infection[Sex == 2], h = 10,
               eval.points = age, display = "none")
estm <- Males$estimate
sem  <- Males$se
estf <- Females$estimate
sef  <- Females$se
plot(Age, Infection, ylab = "Proportion infected", type= "n")
av <- (log(estm/(1-estm)) + log(estf/(1-estf)))/2
se <- sqrt(sem^2 + sef^2)
upper <- 1/(1+exp(-(av + se)))
lower <- 1/(1+exp(-(av - se)))
polygon(c(age, rev(age)), c(upper, rev(lower)),
      col = 6, border = F)
lines(age, estm)
lines(age, estf, lty = 3)
```

6.5 Testing for parallel regression curves

Where there are clear differences between regression curves there are other hypotheses which it may still be of interest to investigate. Figure 6.10 displays data, provided by Ratkowsky (1983), from an experiment designed to investigate the relationship between the yield of onion plants and the density of planting. The two groups refer to two different locations, Purnong Landing (1) and Virginia (2), in South Australia. While it is abundantly clear that onions grown at the Purnong Landing site have systematically higher yields, it is still of interest to know whether the two regression curves might be parallel, corresponding to a difference between the sites of a particularly simple, additive form.

A variety of parametric models have been devised for data of this type. The Holliday model is an example, where the regression relationship for each group is assumed to be of the form $y_{ij} = -\log(\alpha_i + \beta_i x_{ij} + \gamma_i x_{ij}^2)$, with y_{ij} denoting yield on a log scale and x_{ij} denoting density. Analysis of this model suggests that parallel regression curves provide an adequate description of the data. With nonlinear models of this type inferential techniques are no longer exact but are based on linear approximations. Nonparametric techniques offer an alternative means of inference which can be used to check the assumptions of the parametric model.

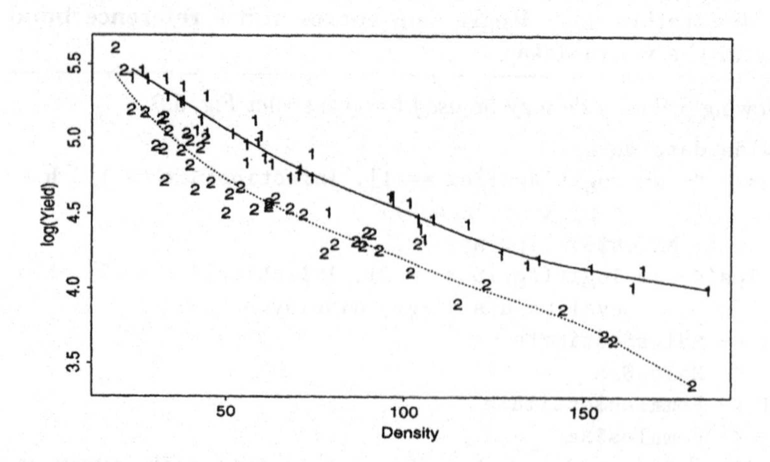

FIG. 6.10. The onions data and two nonparametric regression curves.

In order to reflect an assumption of parallel nonparametric regression lines a model of the form

$$y_{ij} = \alpha_i + m(x_{ij}) + \varepsilon_{ij},$$

where the α_i represent the shifts between the regression lines, can be constructed. In general there may be p lines to compare and so for identifiability, it will be assumed that $\alpha_1 = 0$. In order to fit the model, estimates of these α_i parameters are required. Speckman (1988) shows how this can be done by writing the model in vector-matrix notation as

$$y = D\alpha + m + \varepsilon, \tag{6.3}$$

where α denotes the vector of parameters $(\alpha_1, ..., \alpha_p)$ and D is an appropriate design matrix consisting of 0s and 1s. If α were known then an estimate of m could be constructed of the form

$$\hat{m} = S(y - D\alpha).$$

Substitution of this expression into (6.3) yields the relationship

$$(I - S)y = (I - S)D\alpha + \varepsilon$$

after a little rearrangement. From this the solution

$$\hat{\alpha} = \{D^\top(I - S)^\top(I - S)D\}^{-1}D^\top(I - S)^\top(I - S)y$$

follows by application of the least squares method.

This solution for $\hat{\alpha}$ involves the smoothing matrix S and hence the smoothing parameter h. In order to minimise the bias involved in estimation of α a small

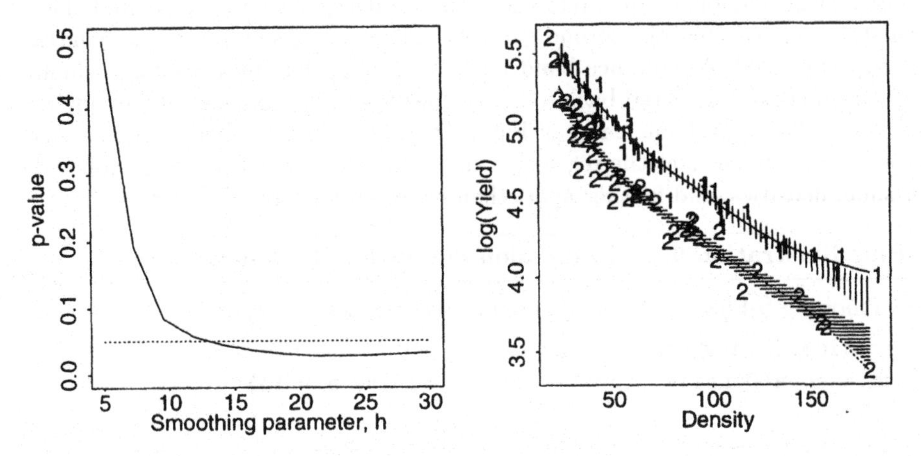

FIG. 6.11. The onions data with reference bands for parallelism.

value of h is recommended for this purpose. A simple guideline is to use the value $2R/n$, where R denotes the range of the design points and n denotes the sample size. With normal kernel functions, this will restrict smoothing to approximately eight neighbouring observations when the data are roughly equally spaced. With the onion data, this produces the estimate $\hat{\alpha}_2 = -0.324$. The effect of different smoothing parameters on the estimation of α_2 is explored in an exercise at the end of the chapter.

If the regression curves are shifted by a distance α_2 then the test of equality discussed in Section 6.4 could be applied. A suitable form of statistic for a test of parallelism is therefore

$$\frac{\sum_{i=1}^{p} \sum_{j=1}^{n_i} \{\hat{\alpha}_i + \hat{m}(x_{ij}) - \hat{m}_i(x_{ij})\}^2}{\hat{\sigma}^2}.$$

The numerator of this statistic can be written as a quadratic form:

$$z^\top \{(I - S_s)D(S_s - S_d)\}^\top (I - S_s)D(S_s - S_d)z,$$

where z denotes the column vector which has $(\hat{\alpha} - \alpha)$ as its first elements and ε thereafter. In pursuing the moment calculations in this case, note has to be taken of the correlation between $\hat{\alpha}$ and ε. This is relatively easy to do because the definition of $\hat{\alpha}$ above is simply a weighted average of the vector of observations y.

The left panel of Fig. 6.11 displays a significance trace for a test of parallelism on the onions data. For a wide range of smoothing parameters the p-value lies a little below 0.05, which casts some doubt on the assumption that the curves are parallel. The fact that the parametric model did not identify this may be due to the less flexible form of the Holliday model which may place some restriction on its ability to capture all the features of these data.

As a graphical follow-up, a reference band for parallelism can be created. The first step is to translate the second curve by a distance α_2 so that the two curves are superimposed. A reference band for equality can then be constructed, and the second curve translated back to its original position, together with a copy of this band. The right hand panel of Fig. 6.11 displays such bands for the onions data. These confirm that the regression curves are closer together than expected for small densities, and further apart than expected for large densities.

S-Plus Illustration 6.7. Regression curves for the onions data.

The following S-Plus *code may be used to reconstruct Fig. 6.10.*

```
provide.data(white.onions)
sm.ancova(Density, log(Yield), Locality, h = 15)
```

S-Plus Illustration 6.8. A significance trace and reference band for parallelism with the onions data.

The following S-Plus *code may be used to reconstruct Fig. 6.11.*

```
provide.data(white.onions)
par(mfrow=c(1,2))
sig.trace(sm.ancova(Density, log(Yield), Locality,
        model = "parallel",display = "none"),
        hvec = seq(5,30, length = 12))
sm.ancova(Density, log(Yield), Locality, h = 15,
        model = "parallel")
par(mfrow=c(1,1))
```

6.6 Further reading

Anderson *et al.* (1994) investigated the properties of a test statistic based on density estimates for comparing two distributions.

Diggle (1990), Diggle *et al.* (1990), Diggle and Rowlingson (1994), Kelsall and Diggle (1995a, 1995b) and Anderson and Titterington (1997) all discuss methods for identifying raised incidence of disease and investigate the laryngeal cancer data in particular. The paper by Kelsall and Diggle (1995a) analyses a larger dataset than the one discussed here.

Speckman (1988), Hall and Hart (1990) and King *et al.* (1991) discuss the comparison of nonparametric regression curves, and Young and Bowman (1995) extend this to nonparametric analysis of covariance.

Exercises

6.1 *Test statistics for equality of densities.* In the discussion of Section 6.2 it was pointed out that a numerical integration approach to the evaluation of a test statistic allows a wide variety of forms to be used. Carry out a test of equality on groups 1 and 2 of the aircraft span data using a test statistic which compares the density estimates on a square root scale, and so is consistent with the reference band.

6.2 *Bandwidths for testing equality of densities.* Since the test of equality of densities discussed in Section 6.2 uses a permutation approach, it is not necessary that the bandwidths used in each density estimate are identical. Carry out a test of equality on groups 1 and 2 of the aircraft span data using a test statistic which compares the density estimates with different bandwidths and observe whether the conclusions of the test are changed.

6.3 *Bias on the square root scale.* Use an appropriate Taylor series expansion to verify that when two densities are compared on the square root scale their principal bias terms remain approximately equal if the same smoothing parameter is used in each estimate.

6.4 *Two-dimensional discrimination.* Exercise 1.8 refers to a two-dimensional discrimination problem. Construct a surface which consists of the standardised difference between the density estimates on a square root scale. Draw the contours of this surface for the values ± 2 and hence construct a variability band for the discrimination curve.

6.5 *Square root scale for density comparisons.* In the discussion of the test carried out in Section 6.3, the idea of using a test statistic of the form $\int\{\sqrt{\hat{f}(y)} - \sqrt{\hat{g}(y)}\}^2 dy$ was discussed, in order to make the variances of the density estimates approximately constant over the sample space. Modify the script for the test of equality on the laryngeal cancer data and observe whether the new test gives a different result.

6.6 *Laryngeal cancer data.* Consider how a nonparametric test which included weighting for the distance from the incinerator could be constructed.

6.7 *Significance trace for the reef data.* Construct a significance trace for the test of equality on the reef data discussed in Section 6.4.

6.8 *Brown onions data.* An additional set of onion data is available in the file `brown.onions`. Carry out a test of equality between the two locations, as with the white onions data. Investigate the effect of the unusually low observation on the evidence for differences between the localities. Construct a significance trace with this observation removed.

6.9 *Follicle data.* These data were amalgamated from different sources, indicated in the `Source` variable. Check that there is no evidence of differences among the sources by carrying out a test of equality on the individual regression curves. Sources 1 and 4 should be omitted because they refer simply to the extreme ends of the data.

6.10 *Comparing logistic curves.* Use a reference band for equality with the low birthweight data, described in Section 3.4, to assist in assessing whether there are differences between the smoking and non-smoking groups.

6.11 *The effect of the smoothing parameter on the estimation of the shifts in the parallel regression model.* The function `sm.ancova` has a parameter h.alpha which allows control over the value of the smoothing parameter used in the estimation of the shift parameters α. Estimate α_2 in the model for the white onions, using a variety of different smoothing parameters. What effect does this have on the parameter estimate and on the assessment of the parallel model?

6.12 *Parallel curves for the reef data.* The reef data have additional catch information expressed in a second score variable which combines information across species in another way. This is available in Score2. Use the ideas discussed in Section 6.4 to identify whether there are differences in catch score 2 between the open and closed zones, using longitude as a covariate. If a difference is detected, assess whether the curves could be assumed to be parallel, so that the effect of preventing trawling might be to increase the log weight for these species by a constant amount at all longitudes. Produce a reference band for parallelism to explore this further.

7

TIME SERIES DATA

7.1 Introduction

The entire discussion so far has been conducted assuming that the data are generated independently of each other. While in many situations one can safely believe that dependence is absent or negligible, it can often be the case that the assumption of independence is untenable.

A typical, quite common situation where dependence is, at least potentially, a prominent feature is represented by time series, and similar data collected over time. In these circumstances, one might well question the appropriateness of the methods discussed in previous chapters. For example, some readers may have noticed that the geyser data used in Section 1.4 fall into this category, and may have objected to the fact that serial correlation has not been taken into account.

In the present chapter, various questions related to serial dependence will be addressed, such as its effect on basic smoothing techniques, and issues more specifically related to the time series context.

There is one classical curve estimation problem in time series which will not be discussed, namely estimation of the spectral density. This has in fact been a long-standing issue in time series analysis, and it involves all the standard problems of density estimation, such as trade-off between bias and variance, from a slightly different perspective. It is therefore not surprising that the kernel method for estimation of a density function was introduced in the statistical literature by two leading time series analysts, Rosenblatt (1956) and Parzen (1962). Since spectral density estimation is such an established topic in time series analysis, the reader is referred to the extensive literature devoted to it; see, for example, Brillinger (1981) and Priestley (1981).

7.2 Density estimation

To focus ideas, it will be useful to refer to the geyser data, introduced in Section 1.4, as a typical example. Figure 7.1 shows an estimate of the marginal density of the duration of eruptions, obtained by applying the methods discussed in Chapters 1 and 2.

Even if this plot is constructed by ignoring the serial nature of the data, it still provides a useful summary of the data. However, there remain some questions which cannot be avoided, such as:

- what can be said about $\hat{f}$ as a formal estimate of the density f, in particular about its bias, variance and consistency?

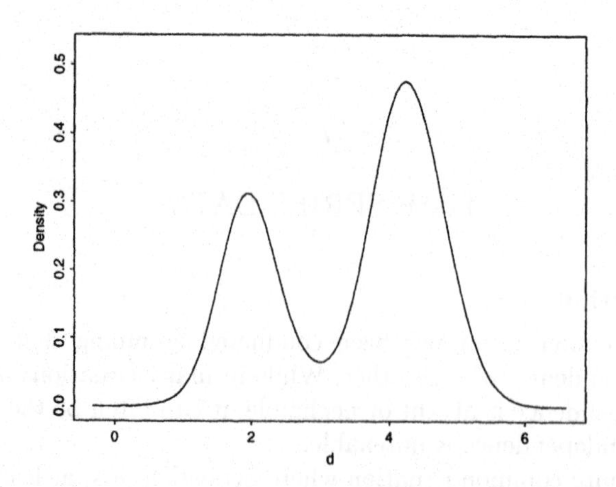

FIG. 7.1. An estimate of the marginal density of the duration of eruptions for the geyser data.

◇ can the methods described in previous chapters, possibly suitably modified, still be used for selecting the bandwidth?

Notice the implicit assumption that the time series is stationary, since otherwise it would not even make sense to talk about f. In particular, stationarity can be safely assumed for a geophysical phenomenon such as the the geyser eruption. Some issues involving non-stationarity will be discussed later in this chapter.

For the bias of $\hat{f}$, the question is easily settled, since the additive nature of (1.1) makes $\text{corr}\{y_i, y_j\}$ (for $i \neq j$) irrelevant, when the expectation is computed; therefore $\mathbb{E}\left\{\hat{f}\right\}$ is exactly the same as for independent data. This conclusion is not surprising at all because it replicates what happens for many other estimation methods, for instance when the sample mean is used to estimate the population mean for dependent data.

Computation of $\text{var}\left\{\hat{f}\right\}$ is a different matter, however. For instance, it is well known that correlation among the data does affect the variance of the sample mean. The good news is that, unlike the sample mean and many other estimators, the variance of $\hat{f}$ is not affected by the correlation, to the first order of approximation (as $n \to \infty$), at least under reasonable assumptions on the convergence to 0 of the autocorrelations when data are increasingly separated in time. Specifically, for n large the approximation

$$\text{var}\left\{\hat{f}(y)\right\} \approx \frac{\alpha(w)}{n\,h} f(y) \tag{7.1}$$

holds, where $\alpha(w) = \int w(u)^2\,du$, and it will be recalled that $h \to 0$ as $n \to \infty$, with $n\,h \to \infty$. Asymptotic normality of $\hat{f}(\cdot)$ is also preserved.

The informal explanation of this convenient fact lies in the local nature of (1.1). More specifically, w operates a form of 'data trimming' which effectively drops a large proportion of the n original data and retains only those close to the selected abscissa y. These 'effective data' are, however, only close on the y axis. In general they are separated on the time axis, hence the correlation among these summands is much reduced, and the situation is quite similar to the case of independent data. To the first order of approximation, the criteria for bandwidth selection are also unaffected, since they are based on considerations of the first two moments.

These conclusions are not limited to the univariate case; similar statements hold for the multidimensional case, when it is necessary to estimate the joint distribution of successive observations. For instance, the joint density of (y_t, y_{t+k}) can be estimated by the methods discussed earlier for bivariate density estimation.

The whole situation is quite comfortable then, since in practice it is legitimate to proceed as if the data were independent, at least for the problem of density estimation. A word of caution is required, however: the whole argument is asymptotic in nature, for $n \to \infty$, and it is possible that, at least for the case of independent data, the convergence of the actual mean and variance to their asymptotic expressions is quite slow. In the present case, the situation is further complicated by the effect of the autocorrelation function, which influences the actual mean and variance in a manner which is difficult to assess. Therefore, it is advisable to resort to the previous asymptotic argument only when n is really quite large, as in fact is the case with the geyser data example.

Figures 7.2 and 7.3 display estimates of the joint density of (y_{t-k}, y_t) for $k = 1$ and $k = 2$, respectively. Inspection of these two figures leads to interesting conclusions on the form of serial dependence of the underlying process.

A Markovian assumption is often adopted in time series as it simplifies both analysis and interpretation of the data. This is particularly true for the case of discrete data, and the present data display some similar characteristics as they are naturally split into 'low' and 'high' values. Figure 7.3 quite distinctly indicates a two-period cyclical behaviour of the process, repeating high and low values if observations are separated by two time units. This pattern is not quite what one would expect from inspection of Fig. 7.2, taking into account the Markovian assumption. In fact, the prominent mode of Fig. 7.2 associated with 'low' values of d_{t-1}, and the comparable height of the other two modes, imply that two similar modes in Fig. 7.3 should be associated with 'low' d_{t-2}. On the other hand, the second plot has two modes with quite different levels in front of 'low' d_{t-2}.

The Markovian assumption is therefore questionable. This conclusion is in agreement with the findings of Azzalini and Bowman (1990), who provide additional evidence, based on a parametric approach, against the first-order Markov assumption.

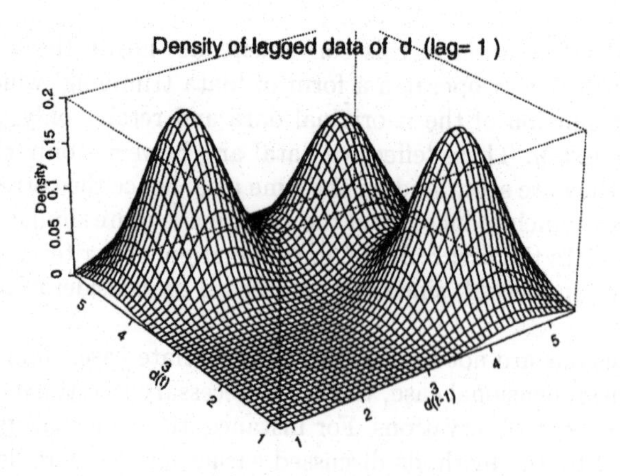

FIG. 7.2. Geyser data: density estimate of (d_{t-1}, d_t).

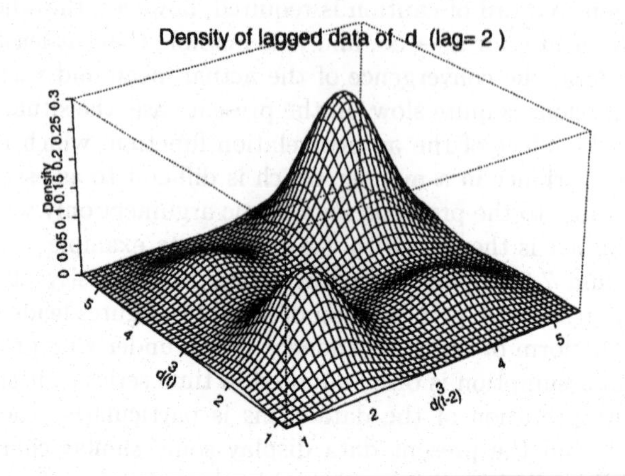

FIG. 7.3. Geyser data: density estimate of (d_{t-2}, d_t).

S-Plus Illustration 7.1. Geyser data

The following S-Plus *code can be used to reconstruct Figs 7.1, 7.2 and 7.3.*

```
d<-geyser$duration
cat("Data are: d=(duration of geyser eruption)\n")
cat("Marginal density of d(t) first, followed by\n")
cat("estimated density of (d(t-k),d(t)), for k=1,2\n")
a<-sm.ts.pdf(d,lags=c(1,2))
```

Mathematical aspects: Mixing conditions and asymptotic results

---

In order for (7.1) to hold, the process $\{y_t\}$ must be such that variables y_{t-k} and y_t behave similarly to independent variables when $k \to \infty$. Properties of this kind, called 'mixing conditions', are standard assumptions in time series analysis in order, for example, to obtain convergence of sample autocovariances to population covariances. The requirement necessary here is therefore no more stringent than usual.

In the literature, there exist various formal specifications of mixing conditions, and correspondingly the method of proving results and their tightness varies somewhat. A simple example is Rosenblatt's 'strong mixing condition', which is in fact quite a mild condition, defined as follows. Suppose that A_t is an event belonging to the time interval $(-\infty, t]$ and B_{t+k} is an event belonging to the time interval $[t+k, \infty)$ with $k > 0$, and denote

$$\alpha(k) = \sup_{A_t, B_{t+k}} \left| \mathbb{P}\{A_t\}\,\mathbb{P}\{B_{t+k}\} - \mathbb{P}\{A_t \cap B_{t+k}\} \right| ,$$

which can be regarded as an indicator of behaviour close to independence for events which are k time units apart. If $\alpha(k) \to 0$ as $k \to \infty$, the process is said to be strong mixing, or also α-mixing. This condition is satisfied by very many 'reasonable' process, such as non-Gaussian autoregressive moving average processes. In the context of interest here, strong mixing is a sufficient condition to establish the results mentioned above, and a number of others.

These results on asymptotic bias, variance and normality of (1.1) can be extended in various directions, which are mentioned briefly, omitting technical details. See Robinson (1983), Györfi *et al.* (1989) and Bosq (1996) for extended discussions.

- If the process $\{Y_t\}$ is d-dimensional and appropriate d-dimensional nonparametric estimation is employed, then convergence to the corresponding multivariate density takes place.
- If the marginal density of $\{Y_t\}$ is estimated simultaneously at d points (of $\mathbb{R}^p$) then convergence is to pd-dimensional multivariate normality, with d independent blocks of p-dimensional components.
- Bandwidth selection via cross-validation (at least in the scalar case) remains an optimal strategy. Specifically, if

$$\mathrm{CV}(h) = \int \hat{f}(y)^2 dy - \frac{2}{n} \sum_{i=1}^{n} \hat{f}^{(-i)}(y_i),$$

where $\hat{f}^{(-i)}(\cdot)$ is the estimate which leaves out y_i, and h^* denotes the minimum of $\mathrm{CV}(h)$, then it can be shown that h^* attains the minimum integrated squared error, $\int (\hat{f}(y) - f(y))^2 dy$, with probability 1 as $n \to \infty$. See Györfi *et al.* (1989, pp. 114–127) for details.

7.3 Conditional densities and conditional means

In time series analysis, the dependence structure of the $\{y_t\}$ process is of primary interest. The conditional distributions given one or more past values are relevant quantities in this setting. These are most naturally estimated as ratios of estimates of joint over marginal densities. By way of illustration, consider the conditional density of y_t given the value taken by the most recent observation, y_{t-1}. It is then quite natural to introduce the estimate

$$\hat{f}_c(y_t|y_{t-1}) = \frac{\hat{f}_2(y_{t-1}, y_t)}{\hat{f}(y_{t-1})},$$

obtained as the ratio between the estimate of the joint density of (y_{t-1}, y_t) and the estimate of the marginal density, considered earlier. For the joint distribution, an estimator similar to the one introduced in Section 1.3 can be used, namely

$$\hat{f}_{1,2}(u, v) = \frac{1}{nh^2} \sum_{t=2}^{n} w_2 \left(\frac{u - y_{t-1}}{h}, \frac{v - y_t}{h} \right).$$

A detailed mathematical discussion of the formal asymptotic properties of this estimate in the current context is given by Robinson (1983); see in particular his Theorem 6.1.

While very informative, conditional densities are not easy to handle, however, since in principle the conditional density of y_t for all possible values of y_{t-1} should be considered. In practice, a certain set of values of y are selected, but this can still be awkward, in general.

In a number of cases, it is then convenient to construct a plot of some summary quantity. For this purpose, the conditional mean function, or autoregression function, is a natural candidate to consider, namely

$$m_k(y) = \mathbb{E}\{Y_t|Y_{t-k} = y\}$$

for $k \geq 1$, or

$$m_{j,k}(y, z) = \mathbb{E}\{Y_t|Y_{t-j} = y,\ Y_{t-k} = z\}$$

for $1 \leq j < k$, or their extensions to several conditional points, which are not explicitly defined here, in order to keep the notation simple. Consider the estimate of the autoregression function $m_k(y)$ given by

$$\hat{m}_k(y) = \frac{\sum_{t=k+1}^{n} y_t\, w(y - y_{t-k}; h)}{\sum_{t=k+1}^{n} w(y - y_{t-k}; h)},$$

which is a direct modification of the Nadaraya–Watson estimate (3.2) with (x_i, y_i) replaced by (y_{t-k}, y_t). This estimator has been studied by a number of authors.

The direct extension of the cross-validation criterion to the present context is given by

$$\mathrm{CV}(h) = n^{-1} \sum_{t=2}^{n} \left\{ y_t - \hat{m}_1^{(-t)}(y_{t-1}) \right\}^2$$

where $\hat{m}_1^{(-t)}(\cdot)$ is the estimate of $m_1(\cdot)$ computed leaving out the pair (y_{t-1}, y_t). Härdle and Vieu (1992) establish the asymptotic optimality of the value of h which minimises $\mathrm{CV}(h)$.

Estimation of $m_k(\cdot)$ or $m_{j,k}(\cdot,\cdot)$ and subsequent plotting is a useful tool for investigating properties such as linearity, at least in an informal way. In addition, these methods can be used in a more formal fashion, by comparing the observed $\hat{m}_k(y)$ curve with some reference regression curve. This idea is used by Azzalini *et al.* (1989) to test whether a general stationary first-order autoregressive model

$$y_t = m(y_{t-1}) + \varepsilon_t\,,$$

where the ε_t are i.i.d. $N(0, \sigma^2)$ variables, can be adequately represented by a linear autoregressive model

$$y_t = \rho y_{t-1} + \varepsilon_t\,.$$

A formal test is developed, based on the PLRT criterion described in Chapter 5. This leads to consideration of the two residual sums of squares

$$\sum_{t=2}^{n} (y_t - \hat{\rho} y_{t-1})^2\,, \quad \sum_{t=2}^{n} (y_t - \hat{m}(y_{t-1}))^2\,,$$

where $\hat{\rho}$ is a moment estimate of ρ and $\hat{m}(\cdot)$ is a minor modification of $\hat{m}_1(\cdot)$, adjusted to decrease bias.

As an alternative to the above estimate of Nadaraya–Watson type, $m_k(y)$ can be estimated by a similar modification of the local linear regression estimate (3.3). The corresponding formal properties have not yet been studied as extensively as those of $\hat{m}_k(y)$; see however Härdle and Tsybakov (1997).

S-Plus Illustration 7.2. Canadian lynx data

The Canadian lynx data consist of the annual number of lynx trappings in the Mackenzie River District of north-west Canada for the period 1821–1934. This famous time series has been analysed by several authors, and it represents a classical benchmark for statistical methods related to time analysis.

The following portion of S-Plus *code illustrates some aspects of the lynx series, with the aid of smoothing methods. The data are log-transformed, as is common practice for these data, because of the extreme skewness and long tail of the original distribution.*

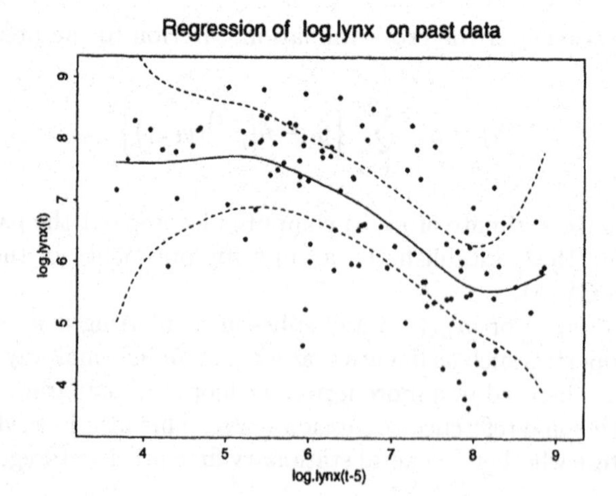

FIG. 7.4. Canadian lynx data: regression function of Y_t on Y_{t-5}.

```
tsplot(lynx)
title("Canadian lynx trapping (1821-1934)")
pause()
cat("Data are now log-trasformed\n")
log.lynx<-log(lynx)
sm.ts.pdf(log.lynx,lags=4:5)
pause()
sm.autoregression(log.lynx,maxlag=5,se=T)
pause()
sm.autoregression(log.lynx,lags=cbind(4,5))
```

The plots produced by this code highlight various features of the lynx data. In particular, there is evidence for the nonlinear nature of the relationship between past and present data. For instance, this aspect is visible in Fig. 7.4 indicating a curvature in the relationship between y_t and y_{t-5}.

Mathematical aspects: Asymptotic variance

Under some regularity conditions in addition to $hn \to \infty$ and the α-mixing condition described in the mathematical remarks of the previous subsection, it can be proved that

$$\frac{\hat{m}_k(y) - m_k(y)}{\sqrt{nh}} \quad \xrightarrow{\text{d}} \quad N\left(0, \alpha(w)\frac{\sigma_k^2(y)}{f(y)}\right)$$

where

$$\sigma_k^2(y) = \text{var}\{Y_t | Y_{t-k} = y\}$$

is the conditional variance given the value of Y_{t-k}. It is worth noticing the similarity between the asymptotic distribution above and the one for independent data and random design points; see, for example, Härdle (1990, p. 100). If the regression curve is estimated at several values of y, then asymptotic normality holds in the multivariate sense, with independence between the components.

7.4 Repeated measurements and longitudinal data

This section and the next deal with situations of a different nature. Suppose now that a reasonable description of the observed data is of the form

$$y_t = m(t) + \varepsilon_t \qquad (t = 1, \ldots, n), \tag{7.2}$$

where $m(\cdot)$ is a deterministic function of time and $\varepsilon_1, \ldots, \varepsilon_n$ are generated by a stationary stochastic process with 0 mean. The function $m(\cdot)$ does not have any specified mathematical form, but it is assumed only to be a smooth function. In other words, (7.2) effectively represents a nonparametric regression setting, except that now the error terms ε_t are serially correlated.

For reasons which will become clear later, it is convenient to start tackling this problem in a repeated measures context, to ease exposition. This means that, instead of a single trajectory, as in (7.2), now N such 'profiles' are available, one for each subject of a group of N homogeneous individuals who share the same function m. The observed data are then of the form

$$y_{it} = m(t) + \varepsilon_{it} \qquad i = 1, \ldots, N;\ t = 1, \ldots, n.$$

An example of data of this type has been reported by Anderson *et al.* (1981), and subsequently used by various authors as a typical example of repeated measures or longitudinal data, as they are also called. Figure 7.5 provides a graphical representation of these data. Here, each profile refers to one of 10 subjects, and for each of them the plasma citrate concentration (μmol l^{-1}) is measured from 8 a.m. for 13 subsequent times at each hour until 9 p.m. The purpose of the analysis is to examine the evolution, if any, of the mean value of the response variable over time.

In this context, data sampled from different individuals can be assumed to behave independently, while data from any single individual are likely to be autocorrelated. Therefore, with the additional assumption that the covariance is stationary, the dependence structure of the ε_{it} is of the form

$$\mathrm{cov}\{y_{it}, y_{kj}\} = \mathrm{cov}\{\varepsilon_{it}, \varepsilon_{kj}\} = \begin{cases} \sigma^2 \rho_{|t-j|} & \text{if } i = k, \\ 0 & \text{if } i \neq k, \end{cases} \tag{7.3}$$

where σ^2 is the variance of the process and $\rho_0 = 1$. If the $n \times n$ covariance matrix of each profile y_i is denoted by V, its entries are

$$\mathrm{cov}\{y_{it}, y_{ij}\} = V_{tj} = \sigma^2 \rho_{|t-j|}\,.$$

In contrast to time series analysis, the 'autocorrelation function' $\{\rho_1, \rho_2, \ldots\}$ is no longer of direct interest. It is simply a nuisance component to be taken into

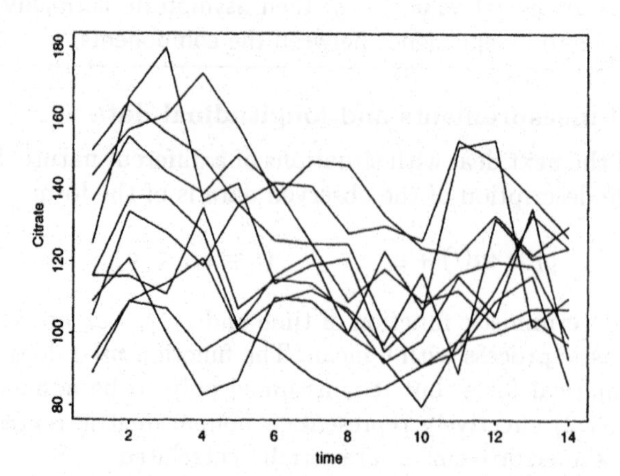

FIG. 7.5. A plot of the individual profiles for the plasma citrate concentration data.

account. This is a reason for not pursuing detailed modelling of the autocorrelation function.

It is most natural to begin by computing the sample averages at each time point, namely

$$\bar{y}_t = \frac{1}{N}\sum_{i=1}^{N} y_{it} \qquad (t = 1, \ldots, n),$$

and estimate $m(\cdot)$ by smoothing the vector of means

$$\bar{y} = (\bar{y}_1, \bar{y}_2, \ldots, \bar{y}_n)^\top \tag{7.4}$$

by some nonparametric regression technique. Notice that the autocorrelation function has been preserved by this averaging operation, since $\operatorname{cov}\{\bar{y}_t, \bar{y}_j\} = \sigma^2 \rho_{|t-j|}/N$.

Denote by $\hat{m}(t)$ the estimate of $m(t)$ for any point of the time axis obtained by smoothing the vector $\bar{y}$ defined by (7.4), and denote by

$$\hat{m} = (\hat{m}_1, \ldots, \hat{m}_n)^\top$$

the vector of values obtained by applying the smoothing procedure when $t = 1, \ldots, n$, and so estimating $m(t)$ at the observed time points.

In a manner similar to that described in Chapter 3, this vector of estimates can be written in matrix form as

$$\hat{m} = S\,\bar{y}, \tag{7.5}$$

where the weights in the $n \times n$ matrix S, which depends on h, form a smoothing matrix.

The bias and variance of this estimate are readily obtained, with little modification from the standard regression situation. Specifically, the mean value and the variance are

$$\mathbb{E}\{\hat{m}\} = S\,m, \qquad \text{var}\{\hat{m}\} = N^{-1} S\, V\, S^{\top},$$

respectively, where $m = (m(1), \ldots, m(n))^{\top}$. Again, the mean value of $\hat{m}$ is not affected by the autocorrelation structure, but var$\{\hat{m}\}$ is, through the matrix V.

The optimal choice of h for minimising the mean squared error of $\hat{m}$ therefore requires knowledge of the autocorrelation function. This brings an additional difficulty. Luckily, in the case of repeated measures data, estimates of the autocorrelations can be obtained by using the residuals

$$e_{it} = y_{it} - \bar{y}_t$$

$(i = 1, \ldots, N;\ t = 1, \ldots, n)$, to estimate the autocorrelation ρ_k by

$$\hat{\rho}_k = \frac{\hat{\gamma}_k}{\hat{\gamma}_0}$$

for $k = 1, 2, \ldots, n-1$, where

$$\hat{\gamma}_k = \frac{1}{nN} \sum_{i=1}^{N} \sum_{t=k+1}^{n} e_{i,t}\, e_{i,t-k}\,,$$

for $k = 0, 1, \ldots, n-1$, is the sample autocovariance at lag k. In this way, an estimate of the autocorrelation function is constructed using the residuals e_{it}, and the estimate of $\hat{m}(\cdot)$ is obtained by smoothing the mean vector $\bar{y}$.

Hart and Wehrly (1986) examine the effect of autocorrelation on mean squared error and on optimal bandwidth selection. They use the Gasser and Müller (1979) estimate but this is not a crucial point of their argument, since the formulae can be written in the linear form (7.5), with different choices of estimate corresponding simply to different choices of the coefficients in S.

Hart and Wehrly (1986) also proposed a criterion for the choice of h in the context of repeated measurements. Denote the residual sum of squares for any specific choice of h by

$$\text{RSS}(h) = \sum_{t=1}^{n} (\bar{y}_t - m_t)^2 = (\bar{y} - \hat{m})^{\top} (\bar{y} - \hat{m}),$$

denote the correlation matrix by

$$R = \sigma^{-2} V = \gamma_0^{-1} V,$$

and denote by $\hat{R}$ the corresponding estimate obtained replacing the γ_k by the $\hat{\gamma}_k$. Hart and Wehrly propose a modification of the Rice (1984) criterion for selecting h, by defining the function

$$M(h) = \frac{\mathrm{RSS}(h)}{n} - \frac{\hat{\gamma}_0}{N}\left(1 - \frac{2}{n}\mathrm{tr}(S\hat{R})\right) \tag{7.6}$$

and choosing the value of h which minimises this. As in other criteria for selecting h, the residual sum of squares is penalised by a quantity depending on the covariance structure of the data. This choice of h is intended to approximate the value of h which minimises

$$\mathrm{MASE}\,(h) = \frac{1}{n}\sum_{t=1}^{n} \mathrm{E}\{\hat{m}_t - m(t)\}^2 .$$

The latter would be the natural 'target function' to consider, but it cannot be evaluated directly, and so $M(h)$ is used as an estimate.

Numerical work by Hart and Wehrly shows that the effect of the autocorrelation on mean squared error and on $M(h)$ can be appreciable, and this effect can lead to higher or lower values of h compared to the value chosen under an assumption of independent data. The relative positions of the two values of h depend on n, N and the ρ_k.

All of the preceding discussion refers to the case of N homogeneous subjects. If there is a subject specific covariate, such as age, which has an effect on each subject's response value, then the technique described above cannot be applied. Some tentative results are given by Raz (1989) and by Azzalini and Bowman (1991), but much has to be done to obtain practically useful methods.

An example of selecting a smoothing parameter for repeated measures data is provided by the citrate data. Figure 7.6 displays the function $M(h)$ for two cases: (i) allowing for correlation among successive data; (ii) assuming independence. The values of h attaining the minimum value of $M(h)$ in the two cases are near 0.8 and 1.2 respectively. Figure 7.7 displays the estimates which result from these two values of smoothing parameter, from which the distinct difference between the two curves is apparent.

A second example is provided by data described by Grizzle and Allen (1969) which are plotted in Fig. 7.8. These refer to a medical study where the response variable is the coronary sinus potassium level measured in dogs at 1, 3, 5, 7, 9, 11 and 13 minutes after coronary occlusion. The aim of this study was to compare different treatments, as is often the case with experiments involving repeated measures. Four treatment groups are present in this case, the first one of which is a control group.

The method described above for smoothing the mean response profile can be applied in the present situation. The modified Rice criterion for the choice of optimal h can be carried out within each group, and then smoothing applied to each mean group profile. Since interest is generally in comparing groups, it is convenient to add variability bands to the mean profile, as an aid to interpretation.

Figure 7.9 shows a comparison between group 3 and the control group. The graph does not show convincing evidence of a difference between the groups.

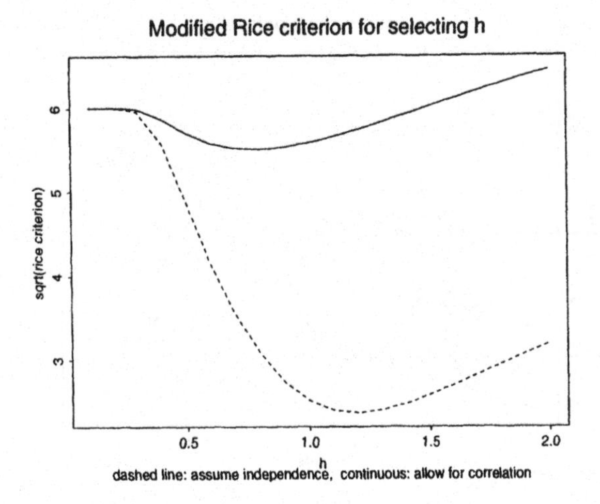

FIG. 7.6. A plot of $M(h)$ versus h for the plasma citrate concentration.

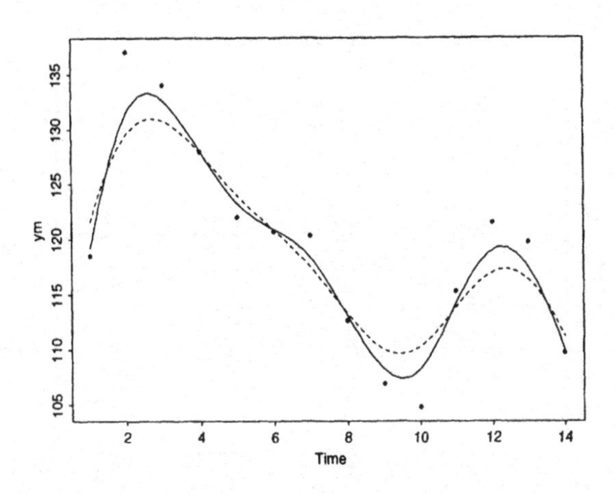

FIG. 7.7. A plot of the raw and the smoothed mean curve for the plasma citrate concentration data. The dashed curve is obtained under an assumption of independence of the observations, while the continuous curve allows for the presence of correlation.

S-Plus Illustration 7.3. Plasma citrate concentration

The following S-Plus *code can be used to reconstruct Figs 7.5, 7.6 and 7.7.*

```
provide.data(citrate)
Citrate<-as.matrix(citrate)
nSubj<-dim(Citrate)[1]
nTime<-dim(Citrate)[2]
```

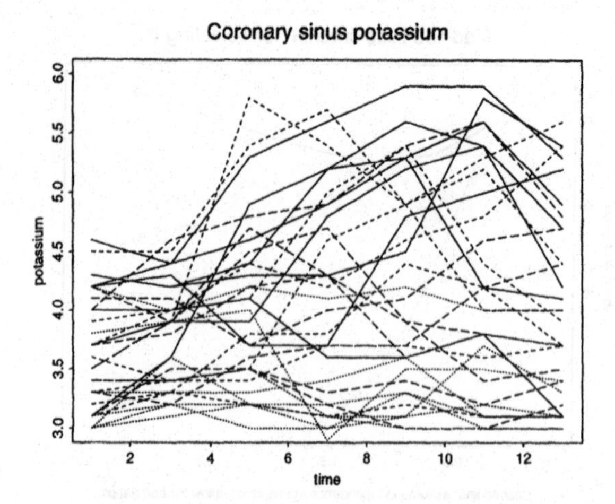

FIG. 7.8. Coronary sinus potassium data: plot of individual profiles (each of the four treatment groups is identified via the line type).

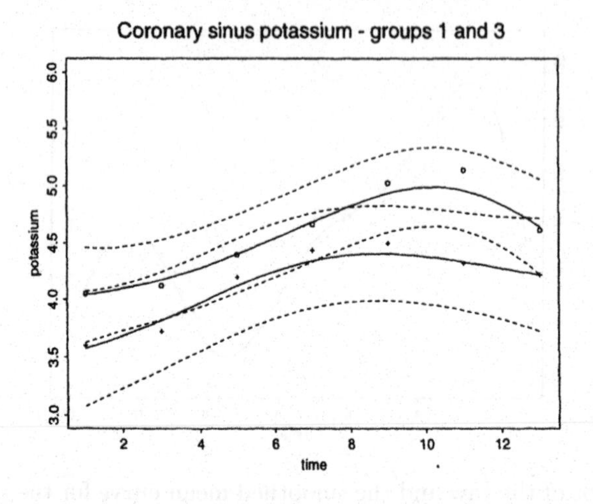

FIG. 7.9. Coronary sinus potassium data: plot of estimated mean profiles and pointwise confidence bands for two groups.

```
Time<-(1:nTime)
plot(c(min(Time),max(Time)), c(min(Citrate),max(Citrate)),
    type="n", xlab="time", ylab="Citrate")
for(i in 1:nSubj) lines(Time,as.vector(Citrate[i,]))
pause()
a <- sm.rm(y=Citrate, display.rice=T)
sm.regression(Time,a$aux$mean,h=1.2,hmult=1,add=T,lty=3)
```

S-Plus Illustration 7.4. Coronary sinus potassium

The following S-Plus code can be used to reconstruct Figs 7.8 and 7.9.

```
provide.data(dogs)
Time <- c(1,3,5,7,9,11,13)
plot(c(1,13), c(3,6), type="n", xlab="time", ylab="potassium")
G <- as.numeric(dogs$Group)
for(i in 1:nrow(dogs))
        lines(Time,as.matrix(dogs[i,2:8]),lty=G[i],col=G[i])
title("Coronary sinus potassium")
cat("\nChoose two groups to compare (Group 1=control)\n")
G     <- ask("First group (1-4)")
id    <- (dogs$Group==G)
Title<- paste(c("Coronary sinus potassium - groups",
          as.character(G)))
gr1   <- as.matrix(dogs[id,2:8])
plot(c(1,13), c(3,6),xlab="time", ylab="potassium", type="n")
sm1   <- sm.rm(Time, gr1, display="se", add=T)
points(Time, sm1$aux$mean, pch=G)
G     <- ask("Second group (1-4)")
id    <- (dogs$Group==G)
gr2   <- as.matrix(dogs[id,2:8])
sm2   <- sm.rm(Time, gr2, display="se", add=T)
points(Time,sm2$aux$mean,pch=G)
Title <- paste(c(Title, "and", as.character(G)), collapse=" ")
title(Title)
```

7.5 Regression with autocorrelated errors

Consider now the case of regression with autocorrelated errors of the form (7.2) when only a single time series is observed. This poses a major difficulty. We already know that, even with independent data, the choice of the degree of smoothing h has a major influence on a nonparametric regression estimate. This difficulty arises from the fact that the most appropriate degree of smoothing should depend on the irregularity of the unknown underlying function $m(\cdot)$, expressed in the second derivative $m''(\cdot)$. Without prior knowledge of $m''(\cdot)$, it is difficult to assess how much of the observed irregularity of the observed data is due to $m(\cdot)$ and how much to $\text{var}\{\varepsilon_t\}$. This is the problem which some data based strategies for selecting h attempt to solve.

If, in addition, autocorrelation is present, the difficulty becomes substantially greater. If the data vary slowly, this may be due to a smoothly varying $m(\cdot)$ with small error variance, or to a constant $m(t)$ with errors which are highly correlated, producing smooth data trajectories. The situation is very unclear, since the data themselves cannot be of much help in deciding between the two alternatives, unless there is some separate source of information either about $m''(\cdot)$ or about the autocorrelation function.

This 'separate source of information' was in fact available in the case of repeated measurements of the previous section. There, the residuals e_{it} could be used to estimate the autocorrelations while the sample means $\bar{y}_t$ were smoothed to estimate $m(\cdot)$. This helpful separation of roles is not possible when $N = 1$, and a single sequence has to be used for estimating both the autocorrelations and $m(\cdot)$.

From this perspective, it is not surprising that automatic procedures, such as cross-validation, for selecting h under the assumption of independence break down completely in the presence of autocorrelation. Hart (1991, pp. 177–178) has shown that, under rather mild conditions, cross-validation tends to choose a value of h which nearly interpolates the data. For this phenomenon to occur it is not necessary to have a high degree of autocorrelation; an AR(1) process with $\rho = 1/4$ may suffice. Hart has provided a formal proof of a phenomenon observed and discussed by a number of authors.

Due to the complexity of the problem, the following discussion is restricted to a relatively simple method discussed by Altman (1990). For a more refined and elaborate approach, see Herrmann *el al.* (1992). In all cases, the methods rely on the assumptions that n is large and that the mean function $m(t)$ is much smoother than the fluctuations of the random component ε_t, despite the fact that serial correlation can produce a smooth pattern in this component.

The main steps of the method of Altman (1990) are as follows.

(a) First, construct a preliminary estimate of $m(\cdot)$, choosing a value of h which will oversmooth rather than undersmooth.

(b) Use the residuals from the first fit to estimate the autocovariances of ε_t, and its corresponding correlation matrix R.

(c) Use these autocovariances in a generalised cross-validation criterion, similar to (7.6), defined as

$$\mathrm{GCV}_d(h) = \frac{\mathrm{RSS}(h)}{(1 - \mathrm{tr}(S\hat{R}))^2},$$

which must be minimised to select a new value of h, and hence to obtain the final estimate $\hat{m}(\cdot)$.

The GCV_d criterion is denoted 'direct' by Altman (1990), in contrast with a variant proposal called 'indirect'. After carrying out steps (a) and (b) as before, the data y are 'whitened' by computing $\tilde{y} = \hat{R}^{-1/2}y$ and a modified 'residual sum of squares'

$$\mathrm{RSS}^*(h) = (\tilde{y} - S\tilde{y})^\top(\tilde{y} - S\tilde{y})$$

used to construct the 'indirect' criterion

$$\mathrm{GCV}_i(h) = \frac{\mathrm{RSS}^*(h)}{(1 - \mathrm{tr}(S))^2}.$$

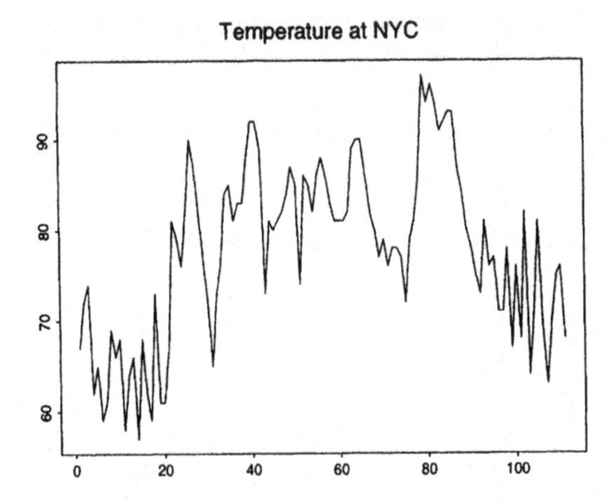

FIG. 7.10. Temperature in New York City: plot of raw data

The two variant methods are similar in motivation, and they also tend to produce similar outcomes.

Since the method involves the correlation matrix R, and hence all terms of the covariance function up to lag $n-1$, it is difficult to obtain a good estimate of R. Use of the sample covariances would produce inconsistent results when the lag of the terms is comparable to n. To overcome this difficulty, a simple solution is to fit a low-order autoregressive process to the series of residuals, and to compute all the remaining terms of the covariance function from the theoretical covariance function of the fitted autoregression.

Figure 7.10 displays a time series of measurements of air temperature in New York City. The presence of some form of dynamics, possibly interpretable as 'non-stationarity', is clear. Figure 7.11 shows the autocorrelation function of the residuals after subtracting a preliminary smooth estimate of the trend in the data. The marked presence of serial correlation is apparent.

Figure 7.12 plots the generalised cross-validation criterion for selecting a smoothing parameter, under an assumption of independence and using the two modified approaches described above. The principal difference among the curves is that ignoring serial dependence leads to a much smaller value of smoothing parameter.

Figure 7.13 shows the estimates corresponding to the choices of h determined by Fig. 7.12. It is clear that ignoring correlation leads to a curve which exhibits too much variation.

S-Plus Illustration 7.5. Air temperature in New York City

The following S-Plus *code can be used to reconstruct Figs 7.10, 7.11, 7.12 and 7.13. In the last two cases this code in fact produces only one curve instead of three. The three different curves correspond to different ways of handling the*

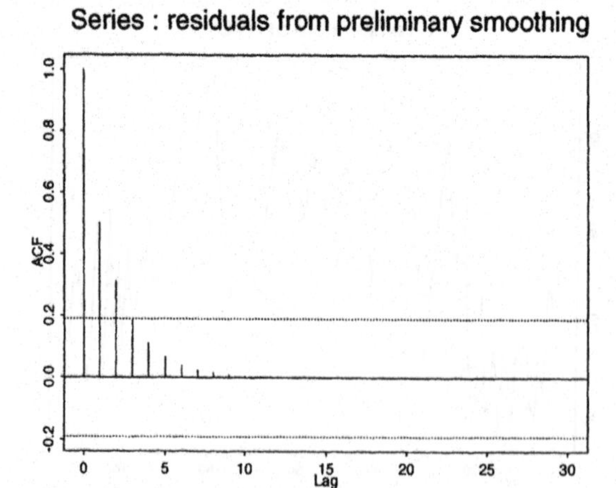

FIG. 7.11. Temperature in New York City: autocorrelation function of the residuals after the preliminary smoothing

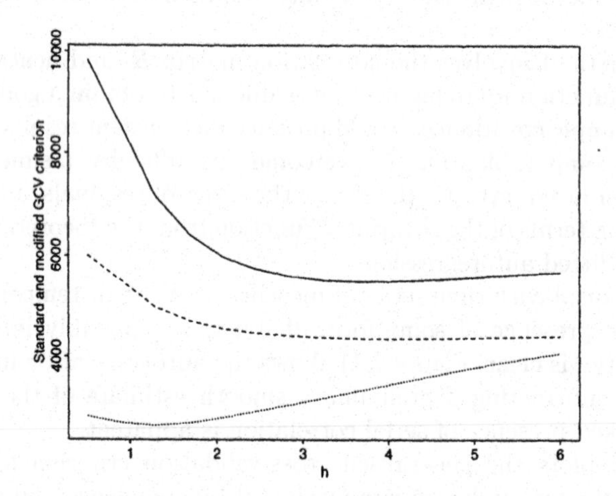

FIG. 7.12. Temperature in New York City: plot of GCV(h) versus h; three variants of the criterion are shown (assuming no correlation, allowing for serial correlation by the 'direct' method, and by the 'indirect' method)

correlation: (i) assuming that there is no serial correlation, (ii) allowing for its presence by the the 'direct' method above, (iii) using the 'indirect' method. These options are controlled by the parameter `method` *of the* `sm.regression.autocor` *function. The possible values of this parameter are* `no.cor`, `direct` *and* `indirect`. *The figures summarise three separate runs of the code, one for each of the three methods. For the current data, and initial smoothing parameter* `h.first=6`, *the three minima occur near* $h = 1.29, 3.64, 4.03$, *respectively.*

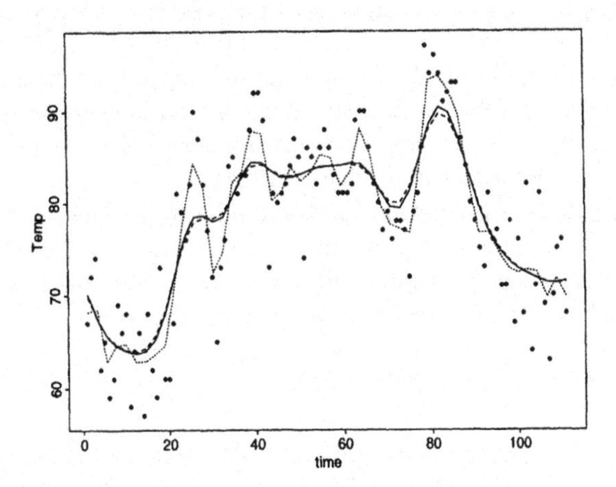

FIG. 7.13. Temperature in New York City: raw data and smoothed curves, obtained by using different criteria for selection of h.

```
Temp<-air$temperature
tsplot(Temp)
title("Temperature at NYC")
sm.regression.autocor(y=Temp, h.first=10, maxh=6)
```

7.6 Further reading

Early results on the properties of the kernel estimator in the time series context are given by Rosenblatt (1971) and later by Robinson (1983). In recent years, the amount of published material in this area has increased enormously, and an exhaustive list is impractical. In addition to the those mentioned in the text, useful results are given by Fan and Gijbels (1996, Chapter 6). Bosq (1996) provides a particularly extensive and mathematically detailed discussion of this topic. The extended review paper by Härdle *et al.* (1997) includes a detailed bibliography.

Roussas (1991) contains a wealth of papers on nonparametric estimation, several of which deal with time series and related problems. Härdle and Tsybakov (1997) develop methods particularly suited for economic time series.

For the analysis of longitudinal data using a nonparametric approach, important pioneering work has been conducted by Gasser *et al.* (1984) and by Müller (1988). Rice and Silverman (1991) present an approach based on penalised least squares for estimation of the mean response curve, leading to a spline formulation. In addition, they develop a method for estimating nonparametrically the covariance structure of the process. In the frequency domain approach, Brillinger (1980) put forward an 'analysis of power table' which provides a decomposition of a spectral density into components, in a manner similar to analysis of variance.

An alternative approach with an analogous purpose is described by Diggle and Al-Wasel (1997).

On the applied side, Shi *et al.* (1996) analyse longitudinal data on HIV infection level by a semiparametric approach where the individual profile is estimated using splines. A semiparametric approach involving kernel smoothing of the trend function is presented by Zeger and Diggle (1994).

The independent S-Plus module for the analysis of repeated measurements data `rm.tools`, summarised by Azzalini and Chiogna (1997), includes simple smoothing facilities of mean profiles. The module is available from the WWW software distributor StatLib under the 'S Archive' chapter.

Exercises

7.1 *Conditional density of geyser data.* Obtain the conditional density function of the duration given that the past value was 'high' (say, 5 minutes). Compare $\hat{f}_c(y_t|y_{t-1}=5)$ with $\hat{f}_c(y_t|y_{t-2}=5)$.

7.2 *Spot the difference.* In Section 7.5, a warning was given about the use of standard criteria for bandwidth selection developed for independent data (such as regular cross-validation) in the present context of time series. However, the message emerging from Sections 7.2 and 7.3 is completely different, encouraging use of standard methods even in those settings. In both cases the data are serially dependent, but there are different prescriptions; what is the cause of the difference?

7.3 *Coronary sinus potassium data.* Use the script provided in the text to compare other groups, similarly to Fig. 7.9. What conclusions can be drawn about differences among groups?

7.4 *Unbalanced data.* Repeated measurements data have been considered in the restricted case of balanced data, forming a matrix of response values (y_{it}). Discuss the problems connected with the extension of the methods to unequal observation times for different subjects. Take into account that very many patterns of unbalancing are possible, ranging from a set of common observation times for the subjects, except for a single missing value in an otherwise complete matrix (y_{it}), to a fully unbalanced situation where all subjects have different observation times.

7.5 *New York City temperature data.* Examine the effect of setting the parameter `method` in the function `sm.regression.autocor()` equal to one of `no.cor`, `direct`, `indirect`. Experiment with different values of the initial smoothing parameter, `h.first`, and examine its effect on the outcome; in doing so, keep in mind that the initial smoothing parameter must not be too small, since undersmoothing must be avoided. The interval which is searched for the final smoothing parameter h is controlled by the parameters `minh` and `maxh`; the number of h values examined is determined by `ngrid`.

7.6 *New York City air data.* The 'data frame' `air`, supplied with S-Plus, contains other time series of New York City air measurements, in addition to the temperature. Modify the S-Plus script of Section 7.5 to handle these other series. Before running the procedures, plot the data in a form like Fig 7.10, and make your guess about the appropriate value of h; then compare this value with the one suggested by `sm.regression.autocor()`, both assuming uncorrelated data and allowing for correlation.

7.7 *Regression under autocorrelation.* The function `sm.regression.autocor()` allows the presence of a `x` variable, assumed to be in increasing order (if not set, `x` is taken to be $1, 2, \ldots, n$, as in S-Plus Illustration 7.5). To try this feature, consider again the 'data frame' `air` and examine for instance the relationship between `temperature` and `ozone` with a command such as

```
sm.regression.autocor(temperature, ozone, h.first=10,
                      minh=1, maxh=15)
```

Repeat this for other plausible pairs of variables in the same dataset. In all cases, try first to make your own guess of the appropriate value of h, and then compare it with that proposed by the procedure.

8

AN INTRODUCTION TO SEMIPARAMETRIC AND ADDITIVE MODELS

8.1 Introduction

In the previous chapters of this book a variety of data structures have been considered. In regression problems, these have usually involved one or two covariates with possible comparisons across groups. However, it is important that any approach to modelling allows considerable generality and flexibility in the data which can be handled. In fact, many of the techniques discussed so far form the building blocks of more general tools for regression, known as semiparametric, additive and generalised additive models. Texts such as Hastie and Tibshirani (1990) and Green and Silverman (1994) provide excellent descriptions of these models. This chapter therefore aims to provide only a brief introduction, principally by illustration, to establish a link with the techniques discussed earlier.

It has been one of the aims of earlier chapters to emphasise inferential techniques. The relatively simple, but none the less important, data structures considered have allowed a variety of tools to be applied effectively. Much has still to be done in the development of inferential methods for the more general models now to be considered. Approximate techniques exist, motivated by analogy with generalised linear models. This area is the subject of current research.

8.2 Additive models

In linear regression modelling, a response variable y is related to covariates, say x_1 and x_2, by the expression

$$y_i = \beta_0 + \beta_1 x_{1i} + \beta_2 x_{2i} + \varepsilon_i, \qquad i = 1, \ldots, n. \tag{8.1}$$

Where an assumption of linearity is untenable, components such as $\beta_1 x_1$ may be replaced by polynomial or nonlinear terms. The idea of an *additive* model, as suggested by Friedman and Stuetzle (1981) and substantially developed by Hastie and Tibshirani (1990), is to allow components of the model to take on nonparametric forms. For example, model (8.1) can be extended by writing

$$y_i = \beta_0 + m_1(x_{1i}) + m_2(x_{2i}) + \varepsilon_i, \qquad i = 1, \ldots, n, \tag{8.2}$$

where m_1 and m_2 denote functions whose shapes are unrestricted, apart from an assumption of smoothness and conditions such as $\sum_{i=1}^{n} m_j(x_{ji}) = 0$ for each j, in order to make the definitions of the functions unique. This provides a helpful generalisation of the usual form of regression function, and a means of modelling data which do not conform to linear assumptions.

For the reef data, which were introduced in Section 3.3, the nonlinear effect of longitude, as a proxy for the changing nature of the sea bed, was confirmed in Section 5.3. It is therefore important to adjust for the known effect of longitude when assessing the importance of other potential covariates. The top left panel of Fig. 8.1 displays an estimate of the two-dimensional regression surface of catch score 1 on the two variables latitude and longitude for the closed zone, from the 1992 survey. The presence of latitude allows changes in a north–south direction, parallel to the coastline, to be examined. The regression surface shows some evidence of a gentle increase in score with latitude.

With only two covariates, a surface of this type can provide useful information on the structure of the data. The difficulty with this approach lies in attempts to extend it to deal with more than two covariates. Not only does it become extremely difficult to provide useful visual information, but the statistical properties of the estimator rapidly deteriorate as, for any given sample size, the data are required to provide an estimate of an arbitrary smooth surface in increasing dimensions. An additive model such as (8.2), which creates an estimate by the combination of a collection of one-dimensional functions, therefore provides an attractive alternative approach. The assumption that the contribution of each covariate is additive is analogous to the assumption made in the usual linear model such as (8.1) and allows much more efficient estimation of each component, and hence of the regression surface.

The additive assumption also has the advantage of allowing many of the well developed one-dimensional methods of estimation to be employed. Firstly, the intercept parameter β_0 can be estimated by the mean of the responses $\bar{y}$, in view of the restrictions on each additive component to sum to zero over the observed covariate values. Then, in order to fit model (8.2) to the reef data, a natural starting point is the estimate of the component m_2 provided by a nonparametric regression of score on longitude (x_2). This estimate can be written in the usual matrix notation as $\hat{m}_2 = S_2(y - \bar{y})$. A rearrangement of model (8.2) as $y_i - \beta_0 - m_2(x_{2i}) = m_1(x_{1i}) + \varepsilon_i$ suggests that an estimate of component m_1 can then be obtained by smoothing the residuals of the data after fitting $\hat{m}_2$,

$$\hat{m}_1 = S_1(y - \bar{y} - \hat{m}_2)$$

and that, similarly, subsequent estimates of m_2 can be obtained as

$$\hat{m}_2 = S_2(y - \bar{y} - \hat{m}_1).$$

At each step, the estimates can be adjusted to sum to zero over the observed design points, in order to fulfil the identifiability condition mentioned above. In fact, Hastie and Tibshirani (1990, Section 5.3.4) show that in the case of two covariates an explicit solution to the system of equations defined above can

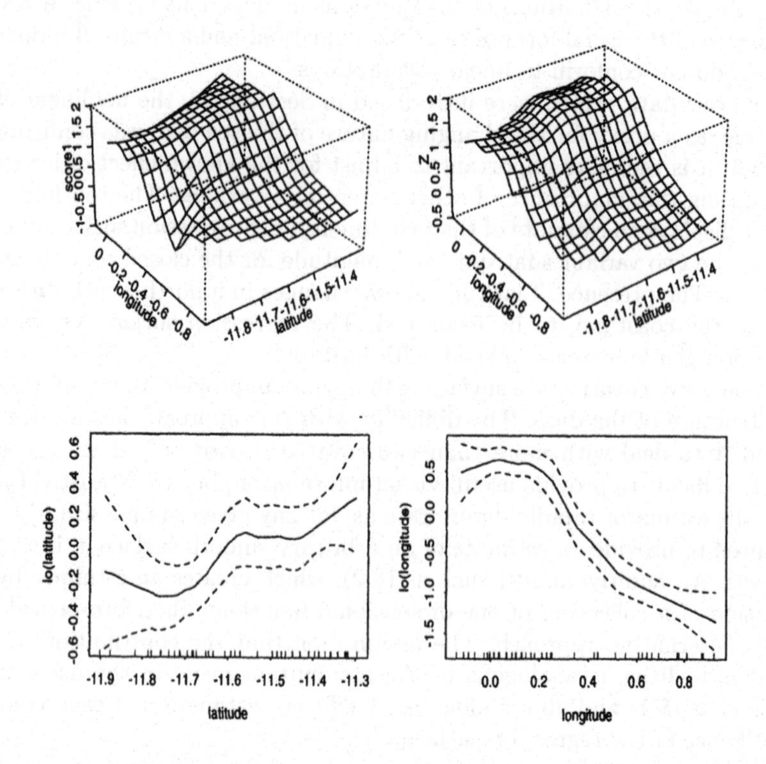

FIG. 8.1. A two-dimensional regression estimate for the reef data using latitude and longitude as covariates (top left panel), and a fitted additive model (top right panel) with its component functions (bottom panels). In order to achieve more attractive scales, 143 has been subtracted from longitude.

be found. However, with a larger number of terms in the model an iterative process, where this converges successfully, will produce joint estimates of each component. This iterative technique, which is in fact an analogue of the Gauss–Seidel method of solving linear least squares regression problems, was called backfitting by Buja *et al.* (1989). It extends easily to models involving several components by repeated estimation of each component through smoothing of the residuals. Specifically, a model of the form

$$y_i = \beta_0 + \sum_{j=1}^{p} m_j(x_{ji}) + \varepsilon_i$$

can be fitted by repeated construction of the smooth estimates

$$\hat{m}_k = S_k \left(y - \bar{y} - \sum_{j \neq k} \hat{m}_j \right).$$

A complete description of the convergence properties of this algorithm is not yet available. Some results do exist, particularly where the smoothing matrices correspond to the use of smoothing splines.

The end result of this process on the reef data is displayed in Fig. 8.1. The full lines in the bottom two panels display estimates of the component functions m_1 and m_2, while the top right panel shows the resulting fitted surface. This surface has the special feature that in each direction the cross-sectional shape across the other direction is identical. Only its location on the vertical scale alters, as determined by the shape of the other component. Clearly there are some functional shapes which additive surfaces of this type will be unable to represent. However, the additive model does seem to capture the principal features of the reef data, and in particular the gentle increase in mean score with latitude is represented. The differences between the two surfaces in Fig. 8.1 may be due simply to random variation.

While models of this type provide very flexible and visually informative descriptions of the data, it is also necessary to consider how models can be compared and inferences drawn. It is not immediately obvious how the inferential techniques discussed in earlier chapters can be extended in a very general way to additive models with several nonparametric components, particularly in view of the iterative nature of the model fitting process. Hastie and Tibshirani (1990) recommend the use of residual sums of squares and their associated approximate degrees of freedom, as described in Section 4.3, to provide guidance for model comparisons. These quantities can be interpreted by analogy with linear models, although the nonparametric nature of additive models means that the usual χ^2 and F distributions do not apply.

For an additive model, the residual sum of squares can easily be defined as

$$\text{RSS} = \sum_{i=1}^{n} (y_i - \hat{y}_i)^2,$$

where $\hat{y}_i$ denotes the fitted value, produced by evaluating the additive model at the observation x_i. Since each step of the backfitting algorithm involves a smoothing matrix S_j, each component of the fitted additive model can be represented in the form $R_j y$, for some matrix R_j, using the notation of Hastie and Tibshirani (1990), who also provide an algorithm for computing an approximate version of these matrices. This allows the calculation of approximate degrees of freedom for each component, and for the entire fitted model, along the lines described for a nonparametric regression model in Section 4.3. In addition, the linear structure of the fitted model also allows standard errors to be computed and variability bands constructed. The details are given in Hastie and Tibshirani (1990, Sections 5.4.4 and 5.4.5).

The choice of the amount of smoothing to apply in the estimation of each component remains an important issue. It would be possible to use an automatic strategy such as cross-validation, but application at every step of the backfitting algorithm would become computationally very intensive. An alternative strategy is to make a subjective specification of the flexibility of the nonparametric components. This can be done conveniently by specifying the approximate degrees of freedom associated with each component, or by specifying the 'span' of the estimator if `loess` is used as the smoothing procedure; see Section 3.6. These are both natural and interpretable scales on which the degree of smoothing can be quantified.

The reef data provide a simple illustration of how model comparisons may be made. There are three obvious models of interest:

Model	RSS	df
1: $\beta_0 + m_1(\text{Latitude}) + m_2(\text{Longitude})$	4.541	33.99
2: $\beta_0 + m_2(\text{Longitude})$	6.128	37.46
3: $\beta_0 + m_1(\text{Latitude})$	27.306	37.54

An assessment can be made of whether there is evidence that a component for latitude should be retained by comparing models 1 and 2. In an obvious notation, this evidence is expressed quantitatively in

$$F = \frac{(\text{RSS}_2 - \text{RSS}_1)/(\text{df}_2 - \text{df}_1)}{\text{RSS}_1/\text{df}_1}, \qquad (8.3)$$

by analogy with the F statistic used to compare linear models. Unfortunately, this analogy does not extend to distributional calculations and no general expression for the distribution of (8.3) is available. However, Hastie and Tibshirani (1990, Sections 3.9 and 6.8) suggest that at least some approximate guidance can be given by referring the observed nonparametric F statistic to an F distribution with $(\text{df}_2 - \text{df}_1)$ and df_1 degrees of freedom. The observed F statistic for the latitude component is

$$\frac{(6.128 - 4.541)/(37.46 - 33.99)}{4.541/33.99} = 3.42.$$

Referring this to an $F_{3.47,33.99}$ distribution produces an approximate p-value of 0.023. This therefore suggests that there is some evidence that the underlying regression surface changes with latitude. The observed F statistic for the longitude component is 48.08 on 3.54 and 33.99 degrees of freedom, which is highly significant, and so confirms the importance of the effect of longitude.

A second example of additive modelling is provided by data on the abundance of mackerel eggs off the coast of north-western Europe, from a multi-country survey in 1992. Watson *et al.* (1992), Priede and Watson (1993) and Priede *et al.* (1995) describe some of the background to these data. Borchers *et al.* (1997) used additive models to describe the effect of covariates and to construct an estimate of the total number of eggs. This can then be used to estimate the total biomass of spawning mackerel by applying a conversion factor based on the number of eggs produced per gram of female fish. Figure 8.2 displays the

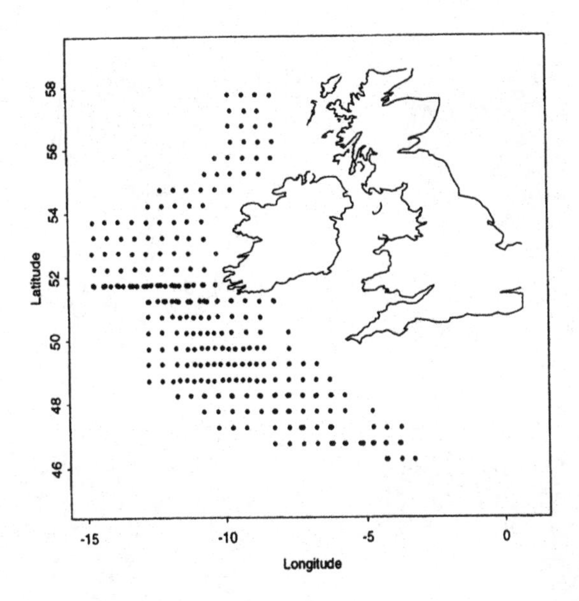

FIG. 8.2. A plot of the sampling points for the mackerel data.

sampling positions of a subset of the data. Further data were collected to the south, in the Bay of Biscay, but these data have rather different features and will be analysed separately in Section 8.4.

Figure 8.3 indicates the pattern of egg densities in the sampled region by using a two-dimensional nonparametric regression estimate with latitude and longitude as covariates. Small values of smoothing parameters have been used in order to display detailed features of the data. A second plot indicates the topography of the sea bed in the same way. The rapid increase in depth, even on a log scale, with distance from the coast is apparent. A plot of density against depth shows that this covariate has a very strong effect, with a preferred depth around 400 to 1000 m, corresponding to the range 6 to 7 on the log scale. On the other hand, there appears to be little relationship between density and sea surface temperature.

An additive model for egg density might reasonably contain terms for depth and temperature, with the latter term present in order to check for effects which may be hidden by a marginal plot. Latitude and longitude can also usefully be included in the model to allow for other geographical effects in the pattern of egg densities. However, it is less natural to include these terms separately since their role is simply to define a two-dimensional co-ordinate system. In contrast to the reef data, the separate co-ordinates cannot be given simple interpretations with respect to the coastline. It is therefore more compelling to introduce latitude and longitude jointly, to reflect spatial position. This leads to the model

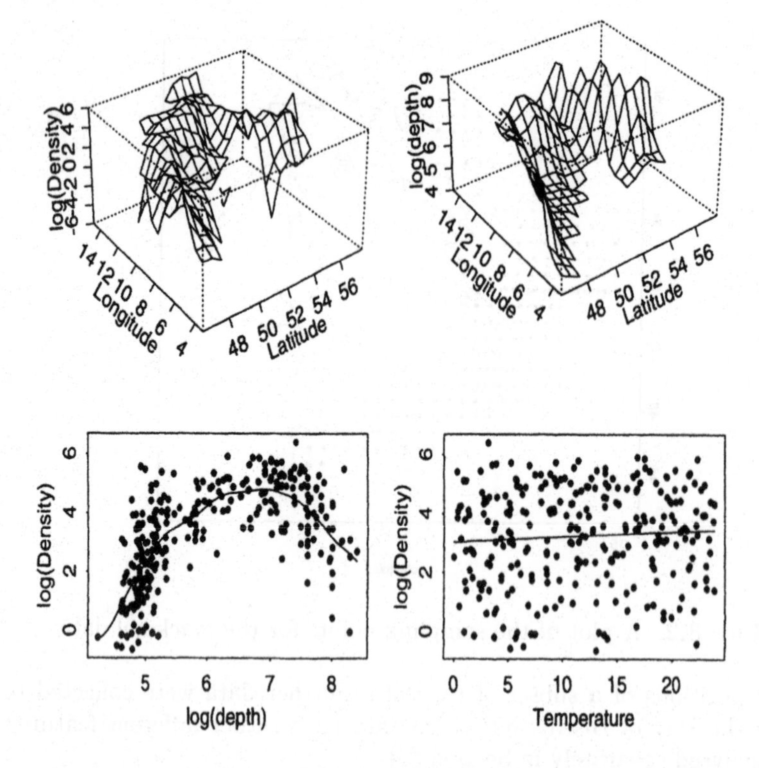

FIG. 8.3. Plots of the covariates for the mackerel data.

$$y = \beta_0 + m_{12}(x_1, x_2) + m_3(x_3) + m_4(x_4) + \varepsilon,$$

where m_{12} represents a smooth two-dimensional function of latitude (x_1) and longitude (x_2), and m_3 and m_4 represent additive terms of the usual type for depth (x_3) and temperature (x_4). Two-dimensional terms require restrictions to define the functions uniquely, as in the one-dimensional case. A simple convention is $\sum_{i=1}^{n} m_{12}(x_{1i}, x_{2i}) = 0$.

The backfitting algorithm described above applies again, since two-dimensional terms can be estimated in the standard form of $S_{12}y$ for a smoothing matrix S_{12}. The following table shows the results of fitting a selection of models. The components of the full model are also displayed in Fig. 8.4.

Model	RSS	df
1: $\beta_0 + m_{12}$(Lat, Long) $+ m_3$(log(Depth)) $+ m_4$(Temp)	261.13	262.80
2: $\beta_0 + m_{12}$(Lat, Long) $+ m_4$(Temp)	360.24	266.51
3: $\beta_0 + m_{12}$(Lat, Long) $+ m_3$(log(Depth))	272.08	266.10
4: $\beta_0 + m_3$(log(Depth)) $+ m_4$(Temp)	335.53	270.99

The large change in residual sum of squares between models 1 and 2 confirms that depth is an important variable. Similarly, the change between models 1 and 4 shows that there are additional geographical effects which should be accounted for in the model by the presence of the term involving latitude and longitude. However, the F value for the temperature effect, namely

$$\frac{(272.08 - 261.13)/(266.10 - 262.80)}{261.13/262.80} = 3.34,$$

when compared to an $F_{3.30,262.80}$ distribution, suggests that the effect of temperature may also be significant. This is portrayed in the top right panel of Fig. 8.4 which shows a gentle increase in egg density with temperature.

Refinements of the component for depth, such as a simple quadratic curve, and the detailed shape of the effect of spatial position might usefully be explored. It might also be wise to consider the possible presence of spatial correlation in the errors. However, after identification of the important variables, one of the aims of this study was to provide an estimate of the total egg biomass over the region. This can be done by constructing the fitted model surface and integrating this over the region of interest, and so the details of the shape of the individual components are less important in this context. Borchers *et al.* (1997) use bootstrapping from the fitted model to construct a standard error of the estimate of the total number of eggs.

S-Plus Illustration 8.1. An additive model for the reef data

The following S-Plus *code may be used to reconstruct Fig. 8.1. Here the* `loess` *estimator is used to construct the smooth component. By default,* S-Plus *uses* `span=0.5`*, which in turn determines the approximate degrees of freedom for each component.*

A substantial section of the code below is devoted to creating a plot of the fitted additive model surface.

```
provide.data(trawl)
ind        <- (Year == 0 & Zone == 1)
score1     <- Score1[ind]
latitude   <- Latitude[ind]
longitude  <- Longitude[ind] - 143
position   <- cbind(latitude, longitude = -longitude)

par(mfrow = c(2,2))
par(cex=0.7)
model1  <- sm.regression(position, score1, h = c(0.1, 0.1))
model2  <- gam(score1 ~ lo(latitude) + lo(longitude))
ex      <- model1$eval.points[,1]
ey      <- model1$eval.points[,2]
ngrid   <- length(ex)
```

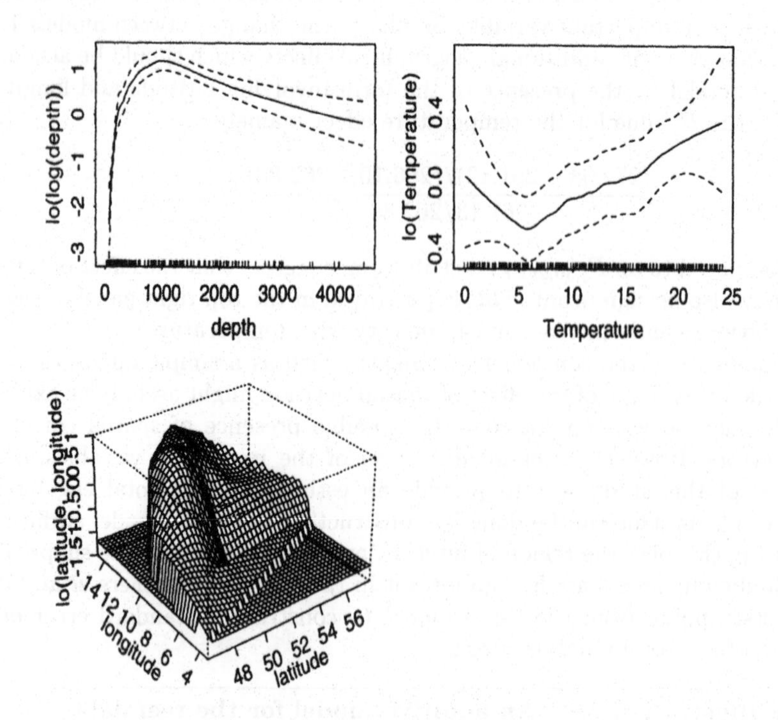

FIG. 8.4. A plot of the components of an additive model for the mackerel data. (Notice that the plot for depth incorporates the log transformation as well as the smoothing procedure into the vertical axis.)

```
grid    <- data.frame(cbind(latitude = rep(ex, ngrid),
             longitude = rep(-ey, rep(ngrid, ngrid))))
surface <- predict(model2, grid)
mask    <- model1$estimate
mask[!is.na(mask)] <- 1
persp(ex, ey, matrix(surface * mask, ncol = ngrid),
             xlab = "latitude", ylab = "longitude")
summary(model2)
plot.gam(model2, se=T)
par(cex=1)
par(mfrow = c(1,1))
```

S-Plus Illustration 8.2. Additive model comparison for the reef data

The following S-Plus code may be used to reproduce the model comparisons discussed in Section 8.2.

```
provide.data(trawl)
ind        <- (Year == 0 & Zone == 1)
score1     <- Score1[ind]
latitude   <- Latitude[ind]
longitude <- Longitude[ind]
print(gam(score1 ~ lo(longitude) + lo(latitude)))
print(gam(score1 ~ lo(longitude)))
print(gam(score1 ~ lo(latitude)))
```

S-Plus Illustration 8.3. Sampling points for the mackerel data

The following S-Plus *code may be used to reproduce Fig. 8.2.*

```
provide.data(mackerel)
plot(-mack.long, mack.lat, xlim=c(-15,1), ylim=c(45,59),
     xlab="Longitude", ylab="Latitude")
britmap()
```

S-Plus Illustration 8.4. Plots of the mackerel data

The following S-Plus *code may be used to reproduce Fig. 8.3.*

```
provide.data(mackerel)
Position  <- cbind(Latitude=mack.lat, Longitude=mack.long)
depth     <- mack.depth
par(mfrow=c(2,2))
sm.regression(Position,     log(Density), h=c(0.3, 0.3), hull=F)
sm.regression(Position,     log(depth),   h=c(0.3, 0.3), hull=F)
sm.regression(log(depth),   log(Density), h = 0.2)
sm.regression(Temperature, log(Density), h = 30)
par(mfrow = c(1,1))
```

S-Plus Illustration 8.5. Additive model comparison for the mackerel data

The following S-Plus *code may be used to reproduce Fig. 8.4.*

```
provide.data(mackerel)
depth      <- mack.depth
latitude   <- mack.lat
longitude <- mack.long
model1   <- gam(log(Density) ~ lo(log(depth)) + lo(Temperature)
              + lo(latitude, longitude))
print(model1)
print(gam(log(Density) ~ lo(Temperature) + lo(latitude, longitude)))
print(gam(log(Density) ~ lo(log(depth))  + lo(latitude, longitude)))
print(gam(log(Density) ~ lo(log(depth))  + lo(Temperature)))
par(mfrow=c(2,2))
```

```
plot(model1, se = T)
par(mfrow=c(1,1))
```

8.3 Semiparametric and varying coefficient models

The generality of additive models of the type (8.2) is attractive. However, precision and power are likely to be lost if a nonparametric component is adopted when a linear or other parametric term is appropriate. There are therefore many situations where a *semiparametric* approach will be beneficial. Models of this type allow mixtures of linear and nonparametric components.

For example, the gentle increase across latitude in the reef data might be modelled quite appropriately by a linear component. If this assumption is correct, such a model will be more powerful in detecting a latitude effect due to the increased precision in parametric estimation. The backfitting algorithm may be used to fit models of this type as before, although a direct least squares solution will be available for all the linear components at each stage of the iteration. A semiparametric model for the reef data of the form

$$y = \beta_0 + \beta_1 x_1 + m_2(x_2) + \varepsilon$$

provides an illustration.

This model is of exactly the same type as the one considered in Section 6.5 and the estimating equations given there remain valid. If the model is written in vector notation as

$$y = D\alpha + m_2 + \varepsilon$$

for a suitable design matrix D and parameters α, then the estimates can be written as

$$\hat{\alpha} = \{D^\top (I - S_2)^\top (I - S_2) D\}^{-1} D^\top (I - S_2)^\top (I - S_2) y,$$
$$\hat{m}_2 = S_2(y - D\hat{\alpha}).$$

Green *et al.* (1985) discussed spline models of this type with agricultural examples. In Section 6.5 it was suggested that a small value of smoothing parameter should be used in the estimation of α. However, the backfitting algorithm uses the same value of h for the estimation of both α and m. This fitted model is displayed in the left panel of Fig. 8.5, from which the combination of linear and nonparametric components is apparent.

Model comparison can again be carried out through F statistics and their associated approximate degrees of freedom . The models of interest for the reef data are as follows:

Model	RSS	df
1: $\beta_0 + m_1(\text{Latitude}) + m_2(\text{Longitude})$	4.541	33.99
2: $\beta_0 + \beta_1\ \text{Latitude} + m_2(\text{Longitude})$	4.956	36.46
3: $\beta_0 + m_2(\text{Longitude})$	6.128	37.46

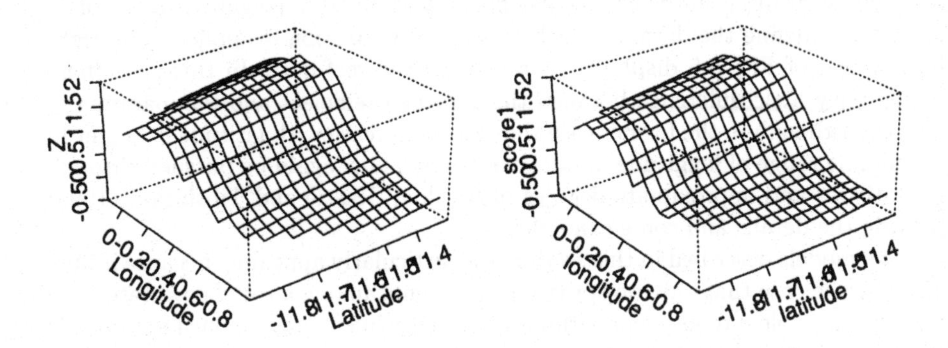

FIG. 8.5. A semiparametric model for the reef data (left panel) and a varying coefficient model (right panel). In each case latitude has a linear effect. In order to achieve more attractive scales, 143 has been subtracted from longitude.

The model comparison of the previous section suggested that there is an effect of latitude. The F statistic for comparing models 1 and 2 is

$$\frac{(4.956 - 4.541)/(36.46 - 33.99)}{4.541/33.99} = 1.26$$

which, when compared to an $F_{2.46,33.99}$ distribution, produces an approximate p-value of 0.30. The semiparametric model 2, with a linear component for latitude, therefore provides an acceptable description of the data. A further comparison of models 2 and 3 confirms the significance of the linear trend in latitude.

Parametric and nonparametric ideas can be combined in a different way to produce a further class of models referred to as varying coefficient models. Here the effect of a covariate is linear, but the coefficients of this linear term may change smoothly with the value of a second covariate. With the reef data, a model such as

$$y = \beta_0(x_2) + \beta_1(x_2)\, x_1 + \varepsilon \qquad (8.4)$$

describes the effect of latitude in a linear manner, but allows the intercept and slope coefficients to vary with longitude. This is an appealing class of models in situations where an underlying parametric relationship is modified by an additional covariate which acts in a nonparametric manner. Hastie and Tibshirani (1993) describe how models of this type can be fitted and used. In the special case of (8.4) above the model can be fitted in a convenient manner by applying two-dimensional nonparametric regression, using the local linear method with a very large smoothing parameter for the x_1 component. This creates an estimator which is linear in x_1 but whose coefficients vary as different neighbourhoods of x_2 are used to define the data to which this linear regression is applied. The right hand panel of Fig. 8.5 displays the result for the reef data. In this case there is relatively little difference between the varying coefficient and semiparametric models, although in the former case the rate of increase with latitude is higher in the middle range of longitude. Model comparison may again be pursued through F statistics and approximate degrees of freedom. Hastie and Tibshirani (1993) provide the details and some examples.

The models described in this section are particularly appealing because it may often be the case that only one or two components among a set of covariates in a regression problem require nonparametric terms, while more standard parametric forms are sufficient for the remainder.

S-Plus Illustration 8.6. Semiparametric and varying coefficient models for the reef data.

The following S-Plus *code may be used to reconstruct Fig. 8.5. A substantial part of the code is devoted to creating a plot of the fitted semiparametric model surface.*

```
provide.data(trawl)
ind        <- (Year == 0 & Zone == 1)
score1     <- Score1[ind]
latitude   <- Latitude[ind]
longitude  <- Longitude[ind] - 143
position   <- cbind(latitude, longitude = -longitude)

model1 <- gam(score1 ~ lo(longitude) + lo(latitude))
model2 <- gam(score1 ~ lo(longitude) +    latitude)
model3 <- gam(score1 ~ lo(longitude))
print(anova(model1))
print(anova(model2, model1))
print(anova(model3, model2))

par(mfrow = c(1,2))
model4   <- sm.regression(position, score1, h = c(0.1, 0.1),
                 display = "none")
ex       <- model4$eval.points[,1]
ey       <- model4$eval.points[,2]
ngrid    <- length(ex)
grid     <- data.frame(cbind(latitude = rep(ex, ngrid),
                 longitude = rep(-ey, rep(ngrid, ngrid))))
surface <- predict(model2, grid)
```

```
mask    <- model4$estimate
mask[!is.na(mask)] <- 1
persp(ex, ey, matrix(surface * mask, ncol = ngrid),
            xlab = "Latitude", ylab = "Longitude")
sm.regression(position, score1, h = c(100, 0.1))
par(mfrow=c(1,1))
```

8.4 Generalised additive models

In Section 3.4 ideas of nonparametric regression were extended to settings where the response variable does not have a normal distribution. Logistic regression for binary or binomial responses is one of the commonest examples of this. The framework of generalised linear models is extremely useful in providing a unified description of a very wide class of parametric models, along with methods of fitting and analysis. In the nonparametric setting a corresponding framework of generalised additive models provides a very flexible form of extension. Excellent overviews of this area are provided in the texts by Hastie and Tibshirani (1990) and Green and Silverman (1994). A brief introduction to the main ideas is provided in this section, based on an illustration.

In discussing the mackerel survey above, it was mentioned that the data collected by Spanish vessels in the Bay of Biscay exhibit rather different features from the remainder of the survey. One of these features is that no eggs were detected at all at a substantial number of the sampling points. This is due to the use of smaller nets and the need to compensate by taking a larger number of smaller volume samples. The sampling positions, together with an indication of the presence or absence of eggs, are displayed in Fig. 8.6. Depth and sea surface temperature are again available as potential covariates. The right panel of Fig. 8.6 shows a local logistic regression curve fitted to the relationship between presence and depth, on a log scale. As in the earlier investigation of the density of eggs, log of depth appears to have an approximately quadratic effect on presence, with an optimal depth around $\exp(6) \approx 400\,\mathrm{m}$.

In the case of logistic regression, a linear model with four covariates takes the form

$$\log\left(\frac{p}{1-p}\right) = \beta_0 + \beta_1 x_1 + \beta_2 x_2 + \beta_3 x_3 + \beta_4 x_4,$$

using the logit transformation defined in Section 3.4. A logistic form of a generalised additive model therefore extends this by replacing each linear component with a nonparametric one. For the Spanish survey data, where the covariates represent latitude, longitude, log of depth and temperature respectively, a natural model is

$$\log\left(\frac{p}{1-p}\right) = \beta_0 + m_{12}(x_1, x_2) + m_3(x_3) + m_4(x_4), \tag{8.5}$$

since latitude and longitude merely define a convenient two-dimensional coordinate system.

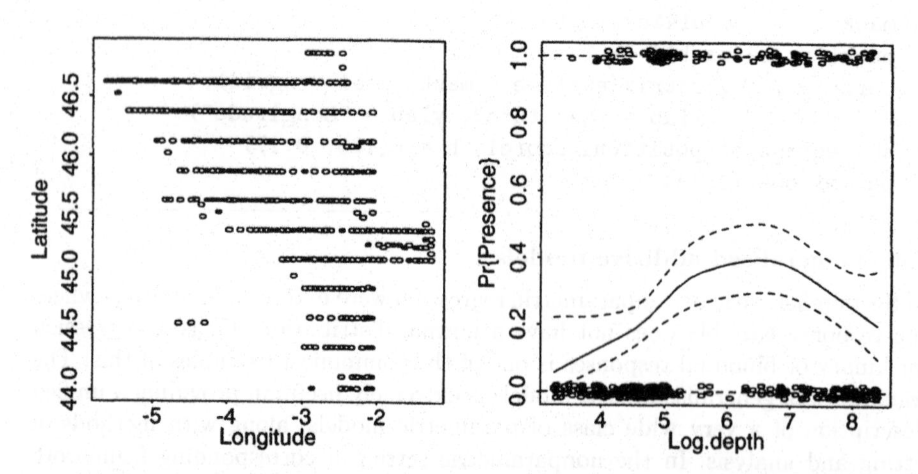

FIG. 8.6. The left panel shows the sampling positions for the Spanish survey with the presence (filled circles) and absence (open circles) of eggs indicated. The right panel shows a local logistic regression estimate of the relationship between presence and log of depth.

In order to fit a generalised additive model the corresponding fitting procedure for generalised linear models again provides a helpful guide. The likelihood function is the natural starting point and a Newton–Raphson, or Fisher scoring, procedure allows the parameter estimates to be located by an iterative algorithm. It is well known that each step of these algorithms can be formulated as a weighted least squares linear regression. This immediately provides a natural analogue in the generalised additive model setting, by employing a weighted nonparametric regression at each step. Hastie and Tibshirani (1990) and Green and Silverman (1994) provide detailed discussion of this approach.

Figure 8.7 displays in graphical terms the results of fitting the generalised additive model (8.5) to the Spanish survey data. The expected concave effect of log of depth is apparent. A more complex nonlinear curve describes the effect of temperature and a two-dimensional function of latitude and longitude captures the additional spatial variation. In order to explore which of these terms can be regarded as evidence of an underlying systematic effect, rather than random variation, different models for the data can be compared. A quantitative measure of the goodness-of-fit of a generalised linear model is available in the deviance, which is the difference of the log-likelihood at the 'saturated' and fitted models. A deviance can also be defined in an analogous manner for a generalised additive model .

For the Spanish survey data, the deviances for a number of models of interest are shown below:

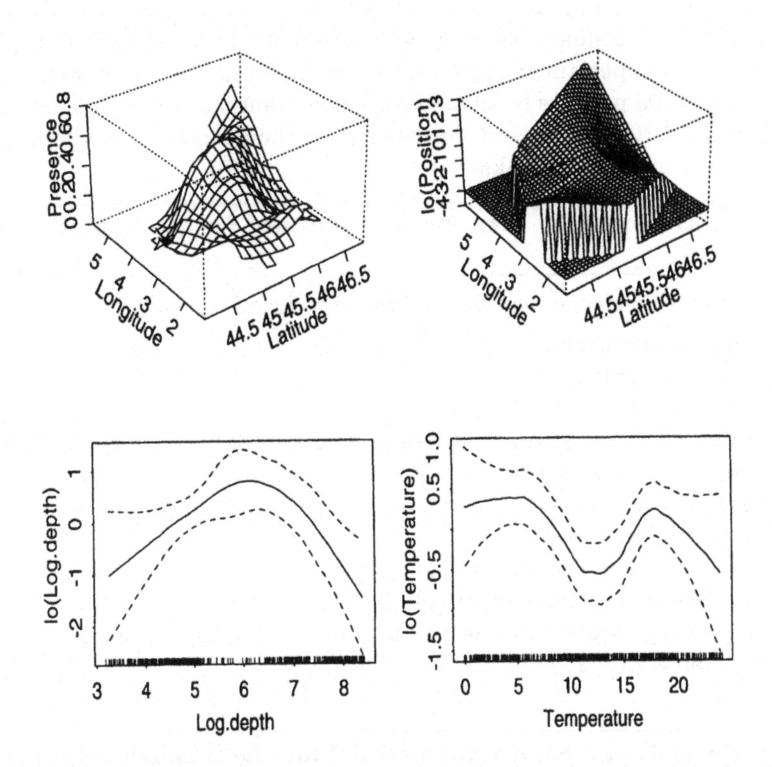

FIG. 8.7. A two-dimensional regression estimate for the probability of the presence of eggs in the Spanish survey data using latitude and longitude as covariates (top left panel). The other panels show the terms of a fitted additive model.

Model	Deviance	df
1: $\beta_0 + m_{12}(\text{Lat, Long}) + m_3(\log(\text{Depth})) + m_4(\text{Temp})$	384.22	401.30
2: $\beta_0 + m_{12}(\text{Lat, Long}) + m_4(\text{Temp})$	394.32	404.75
3: $\beta_0 + m_{12}(\text{Lat, Long}) + m_3(\log(\text{Depth}))$	398.43	404.68
4: $\beta_0 + m_3(\log(\text{Depth})) + m_4(\text{Temp})$	431.95	409.20

As with additive models for data with normal responses, general distribution theory for model comparisons is not available, even asymptotically. However, by applying suitable quadratic approximations, degrees of freedom can be associated with each model, and so some guidance on model comparisons can be taken by comparing differences in deviance to χ^2 distributions indexed by the difference in the approximate degrees of freedom.

An analysis of this type suggests that all of the three components in model 1 contribute significantly. Once again, the main aim in the present context is to construct a model which can be used to amalgamate the egg densities into an estimate of the fish biomass, and so identification of the exact nature of each covariate effect is secondary. The suggested shape of each component is displayed in Fig. 8.7. The temperature effect looks rather implausible from a scientific point of view and should perhaps be treated with some caution. Hastie and Tibshirani (1990, Section 6.10) provide some discussion on the dangers of overinterpreting additive fits in models of this kind.

S-Plus Illustration 8.7. Presence and absence for the Spanish survey data

The following S-Plus *code may be used to reproduce Fig. 8.6.*

```
provide.data(smacker)
Presence <- Density
Presence[Presence > 0] <- 1
Position <- cbind(Longitude=-smack.long, Latitude=smack.lat)
Log.depth <- log(Depth)
par(mfrow = c(1,2))
plot(Position, type="n")
points(Position[Presence==1,], pch=16)
points(Position[Presence==0,], pch=1)
sm.logit(Log.depth, Presence, h = 0.7, display = "se")
par(mfrow = c(1,1))
```

S-Plus Illustration 8.8. Additive model for the Spanish survey data

The following S-Plus *code may be used to reproduce Fig. 8.7.*

```
provide.data(smacker)
Presence <- Density
Presence[Presence > 0] <- 1
Position <- cbind(Latitude=smack.lat, Longitude=smack.long)
Log.depth <- log(Depth)
model1 <- gam(Presence ~ lo(Position) + lo(Log.depth)
                + lo(Temperature), family = binomial)
model2 <- gam(Presence ~ lo(Position) + lo(Temperature),
                family = binomial)
model3 <- gam(Presence ~ lo(Position) + lo(Log.depth),
                family = binomial)
model4 <- gam(Presence ~ lo(Log.depth) + lo(Temperature),
                family = binomial)
print(anova(model1))
par(mfrow=c(2,2))
sm.regression(Position, Presence, h=c(0.3, 0.3), poly.index=0,
                zlim = c(0,0.8))
plot(model1, se = T)
```

```
par(mfrow=c(1,1))
```

8.5 Further reading

Borchers *et al.* (1997) extend the range of models considered in this chapter by adopting an overdispersed Poisson distribution for the error component in analysing the mackerel data.

Hastie and Tibshirani (1990), and more recently Green and Silverman (1994), provide an extensive description and discussion of generalised additive models. Cleveland *et al.* (1992) give very helpful material on the practical aspects of local regression models. These authors, along with Hastie and Tibshirabi (1990), also describe a modified form for the approximate distribution of F statistics.

Green (1987) reviews the penalised likelihood approach to semiparametric regression models.

Buja *et al.* (1986; 1989) apply the concept of nonparametric models and additivity to principal curves in a representation of multivariate data.

From a more parametric point of view, Royston and Altman (1994) describe families of fractional polynomials which can provide flexible regression functions but which also aim to retain some of the properties of parametric models.

Friedman and Stuetzle (1981) introduced the concept of projection pursuit regression, based on the model $y = \beta_0 + \sum_{j=1}^{J} m_j(\beta_j^\top x) + \varepsilon$, which involves nonparametric functions of linear combinations $\beta_j^\top x$ of the vector of covariates x. This allows some further reduction in the dimensionality of the space in which nonparametric estimation is carried out. A further dimension reduction approach, known as sliced inverse regression was introduced by Li (1991) and Duan and Li (1991).

A modification of the additive model is to allow simultaneous selection of a smooth transformation of the response variable y. This is the basis of the alternating conditional expectation and additivity and variance stabilising transformation approaches of Breiman and Friedman (1985) and Tibshirani (1988).

Friedman (1991) introduced a further generalisation, using the idea of tensor-products to create multivariate adaptive regression splines.

Exercises

8.1 *The backfitting algorithm.* Write S-Plus code to implement the backfitting algorithm described in Section 8.2 for the reef data. Use the function `sm.regression` as the means of one-dimensional smoothing at each step. Inspect the result for any major differences from the results produced by Illustration 8.1, which uses the S-Plus nonparametric regression function `loess`.

8.2 *The effect of depth in the mackerel data.* In the explorations of the mackerel data carried out in this chapter, the effect of depth appeared to be approximately quadratic, on a log scale. Fit and compare suitable models to examine whether a quadratic would provide an acceptable description of the data.

8.3 *Comparing generalised additive models and other procedures.* The analysis carried out for many of the examples earlier in this book could be approximated by analysis of a suitable generalised additive model. Investigate the extent to which these analyses agree in the following cases.

- ◇ With the Great Barrier Reef data, examine the evidence for nonlinearity of the effect of longitude on catch score (Chapter 5), compare nonparametric regressions of catch score on longitude for the open and closed zones (Chapter 6) and compare regression surfaces of catch score on latitude and longitude between the two survey years for the closed zone (Chapter 6).
- ◇ With the worm data, compare the regressions of proportion infected against age for the males and females (Chapter 6).
- ◇ With the low birthweight data, compare the regressions of the probability of low birthweight on the mother's weight for the smoking and non-smoking groups (Chapter 6).

8.4 *Ethanol data.* A dataset from an experiment which involved the burning of ethanol in a car engine is supplied with S-Plus in the data frame `ethanol`. Details of the background to the experiment can be obtained by giving the command `help(ethanol)`. Construct a suitable model to describe these data, using smooth additive terms where appropriate. These data were originally reported by Brinkman (1981) and were analysed by Cleveland *et al.* (1992) and Hastie and Tibshirani (1993).

8.5 *Great Barrier Reef data.* In Section 4.6 the effect of depth on the catch score was explored. Use an appropriate additive model to assess whether there is any effect of depth after adjusting for longitude.

APPENDIX A

SOFTWARE

Practical application of the techniques described in this book requires a suitable statistical computing environment. S-Plus is a natural choice because it is widely used within the statistical community and it has a highly flexible structure which allows new procedures to be added to its basic functionality. S-Plus already provides a number of nonparametric smoothing facilities. However, the techniques described in this book required access to the low level operations of smoothing procedures and so it was natural to create a new set of tools within the S-Plus environment. These tools, which together form the `sm` library, are intended to be used in exploring the ideas and techniques described in the text. In particular, S-Plus scripts to reproduce the figures and analyses provided is supplied at the end of most sections.

The `sm` library, and the datasets and scripts used throughout the book, are all available in electronic form. These can be obtained from the following World Wide Web sites:

`http://www.stats.gla.ac.uk/~adrian/sm`
`http://www.stat.unipd.it/dip/homes/azzalini/SW/Splus/sm`

and from the following ftp sites:

`ftp.stats.gla.ac.uk/pub/sm`
`ftp.stat.unipd.it/users/azzalini/sm`

Details on how to install the library are also provided.

The principal role of the functions is to illustrate the ideas described in the text and to allow readers to explore these on real data. Effort has been expended on making the functions reliable, but they are not guaranteed to work under all circumstances. In particular, many parts of the code have been written with more emphasis on readability than computational efficiency. This is particularly true of memory usage, where the large matrices employed in some functions mean that large datasets cannot be used. The exact limits will, of course, depend on the computer being used.

Summary of main functions

The aim of the descriptions provided below is to provide a simple guide to the main parameters of each function. Further details are provided in the on-line help. Only the principal parameters are listed, with dots used to indicate that additional parameters may be used. In almost all functions the first one or two parameters define the data to be used. Other arguments are optional and should be used with their associated keyword since the descriptions below refer only to a subset of the parameters and so they cannot always be passed by order. Facilities for missing data are not provided in any functions.

It may be helpful to indicate the chapters where each function is principally used.

Chapter 1	`sm.density, sm.sphere, nnbr`
Chapter 2	`hnorm, hsj, hcv, sj, cv, nise, nmise`
Chapter 3	`sm.regression, sm.logit, sm.poisson, sm.survival, nnbr`
Chapter 4	`sm.sigma, hcv`
Chapter 5	`sm.logit.bootstrap, sm.poisson.bootstrap, sig.trace`
Chapter 6	`sm.ancova`
Chapter 7	`sm.autoregression, sm.regression.autocor, sm.ts.pdf, sm.rm`

The functions `provide.data` are used throughout to provide access to data. The function `pause` is occasionally used to control the appearance of successive plots. The `gam` function for generalised additive models used in Chapter 8 is a standard S-Plus function whose operation is described in the standard on-line help.

`binning(x, breaks, nbins)`

This function is usable even independently of smoothing techniques. Given a vector or matrix `x`, it constructs a frequency table associated to appropriate intervals covering the `x` range. The user can set either the parameter `breaks` or `nbins` or none of them. In the two-dimensional case, `breaks` must be a matrix with two columns, if supplied.

`cv(x, h, ...)`

This function evaluates the criterion used to determine the smoothing parameter for density estimation defined by cross-validation, using data `x` in one or two dimensions.

`hcv(x, y=NA, ...)`

This function evaluates the smoothing parameter defined by cross-validation. The default value of `y` is missing, in which case density estimation will be assumed, where the data `x` may be one- or two-dimensional. When `y` is present the local linear form of nonparametric regression with the single covariate `x` will be assumed.

`hnorm(x, ...)`

This function evaluates a normal optimal smoothing parameter for a density estimate in one, two or three dimensions. The values returned scale the smoothing parameter which is appropriate for a standard normal distribution by the sample standard deviation of each dimension of the data `x`.

`hsj(x, ...)`

This function evaluates the smoothing parameter for density estimation defined by the Sheather–Jones 'plug-in' approach for univariate data `x`.

`nise(x)`

This function evaluates the integrated squared error between a density estimate constructed from a standardised version of the univariate data `x` and a standard normal distribution.

`nmise(sd, n, h)`

This function evaluates the mean integrated squared error between a density estimate constructed from normal data, with mean zero and standard deviation `sd`, and a normal density with the same parameters.

`nnbr(x, k, ...)`

This function returns the `k`th nearest neighbour distances from each element of `x` to the remainder of the data. `x` may be one- or two-dimensional.

`pause`

This function causes execution of S-Plus to halt until a key is pressed. This is used in some scripts to allow the user control over the rate of appearance of successive plots.

`provide.data(name)`

This function provides access to the dataset identified by `name`. For flexibility, the datasets are provided in ascii form, with the names of each variable listed in the first row of the file. This function reads the files and makes the data available as a data frame.

`sj(x, ...)`

This function evaluates the criterion used to determine the smoothing parameter for density estimation defined by the Sheather–Jones 'plug-in' approach for univariate data `x`.

`sigtrace(expn, hvec, ...)`

This function produces a significance trace for a hypothesis test. The S-Plus expression `expn` should define the hypothesis test required but should not include a value the smoothing parameter `h`. The test is applied over the range of smoothing parameters defined by `hvec` and the p-values produced are plotted against these smoothing parameters.

`sm.ancova(x, y, group, h, model, h.alpha, ...)`

This function carries out a nonparametric analysis of covariance by comparing the regression curve in `x` and `y` over the different sets of data defined by `group`. In addition to the parameter `h` which controls the degree of smoothing across `x`, an additional parameter `h.alpha` controls the degree of smoothing which is applied when the curves are assumed to be parallel and the vertical separations are estimated. When the `model` parameter is set to `"equal"` or `"parallel"` a test of the appropriate model is carried out and a reference band is added to the plot.

`sm.autoregression(x, h, lags, ...)`

This function estimates the conditional mean value of the time series `x`, assumed to be stationary, given its past values. If the parameter `lags` is a vector, then the function estimates $\mathbb{E}\{x_t|x_{t-k}\}$ for all values of k in the vector `lags`. If `lags` is a matrix with two columns, then it estimates $\mathbb{E}\{x_t|x_{t-k}, x_{t-r}\}$ for each pair (k, r) appearing in a row of `lags`.

`sm.density(x, h, h.weights, model, display, panel, positive, ...)`

This function creates a density estimate from the data in `x`, which are assumed to be on a continuous scale. `x` can be either a vector or a matrix with two or three columns, describing data in one, two or three dimensions. The smoothing parameter `h` should be a vector whose dimensionality matches that of `x`. If this argument is omitted, normal optimal smoothing parameters are used. Weights to define variable smoothing parameters can be applied through the vector `h.weights`, whose length should match the number of observations. The parameter setting `model="normal"` adds reference bands for a normal model. The effect of the `display` parameter depends on the dimensionality of `x`. In one dimensions variability bands can be created through `display="se"`. In two dimensions, three different representations of the estimate are available through the settings `"persp"`, `"image"` and `"slice"`. The parameter setting `panel=T` allows some aspects of the function to be controlled through an interactive menu. For one-dimensional data this panel also allows the effect of the smoothing parameter on the shape of the estimate to be viewed as an animation. For data in one or two dimensions, `positive=T` invokes a transformation approach which is appropriate when the data take only positive values and where boundary problems may occur near the origin.

`sm.density.compare(x, group, model...)`

This function allows two or more univariate densities estimates to be compared. The default value of the `model` parameter is `"none"`, in which case the densities are simply drawn on a common set of axes. If `model` is set to `"equal"` then a bootstrap test of equality is carried out. If the comparison involves only two density function the the test will be accompanied by a graphical reference band for equality.

`sm.logit(x, y, N, h, ...)`

This function estimates the regression curve using the local likelihood method, assuming that `x` is a vector with values of a covariate and `y` is a vector of associated binomial observations. If the parameter `N` is present, then it represents the vector of binomial denominators, otherwise it is assumed to consist entirely of 1's.

`sm.logit.bootstrap(x, y, N, h, degree, ...)`

This is associated with `sm.logit()` for the underlying fitting procedure. It performs a PLRT for the goodness-of-fit of a standard parametric logistic regression model. The optional parameter `degree` (normally set equal to

1) regulates the degree of the polynomial in `x` appearing in the linear predictor.

`sm.poisson(x, y, h, ...)`

This function estimates the regression curve using the local likelihood method, assuming that `x` is a vector with values of a covariate and `y` is a vector of associated Poisson observations.

`sm.poisson.bootstrap(x, y, h, degree, ...)`

This is associated with `sm.poisson()` for the underlying fitting procedure. It performs a PLRT for the goodness-of-fit of a standard parametric loglinear regression model. The optional parameter `degree` (normally set equal to 1) regulates the degree of the polynomial in `x` appearing in the linear predictor.

`sm.regression(x, y, h, h.weights, model, display, panel, ...)`

This function constructs an estimate of the regression curve of `y` on the one- or two-dimensional covariate `x` using nonparametric smoothing. The default settings use a local linear regression approach. Weights to define variable smoothing parameters can be applied through the vector `h.weights`, whose length should match the number of observations. The parameter settings `model="no effect"` or `"linear"` carry out tests of the appropriate model and construct reference bands. In one dimension variability bands can be created through `display="se"`. The parameter setting `panel=T` allows some aspects of the function to be controlled through an interactive menu. For one-dimensional data this panel also allows the effect of the smoothing parameter on the shape of the estimate to be viewed as an animation.

`sm.regression.autocor(x, y, h.first, ...)`

This function estimates the regression function of `y` on `x` when the error terms are serially correlated. The appropriate smoothing parameter is selected using a suitable procedure which requires a preliminary selected value `h.first`, which should tend to oversmooth. The parameter `x` can be omitted, and in this case it is assumed to be $1, 2, \ldots, n$.

`sm.rm(Time, y, ...)`

Assume that the matrix `y` contains repeated measurements, observed on n subjects, each of them observed at time points `Time`. The appropriate smoothing parameter is selected allowing for the presence of serial correlation within a given subject, and the mean profile is estimated.

`sm.sigma(x)`

This function estimates the standard deviation of the errors in a nonparametric regression model with one covariate `y`.

`sm.sphere(lat, long, phi, theta, kappa, ...)`

This function plots a density estimate for directional data which can be represented as points on the surface of a sphere. The locations of the data as indicated by the latitude and longitude vectors `lat` and `long`. The angles phi and theta determine the viewing angle of the sphere. The parameter kappa is the smoothing parameter, which is expressed as the scale parameter of a Fisher distribution.

sm.survival(x, y, status, h, hv, p, ...)
This function constructs p percentiles of censored survival data y as a smooth function of a covariate x. status indicates whether each observation is censored. By default, uncensored observations are represented by 1. In addition to the parameter h which controls the degree of smoothing across the covariate x, as additional parameter hv controls the degree of smoothing of the weighted Kaplan–Meier estimator from which the percentiles are constructed. hv refers to a 0–1 scale and takes the default value of 0.05.

sm.ts.pdf(x, lags, ...)
This function estimates the density function of the time series x, assumed to be stationary. The marginal density is estimated in all cases; the bivariate density of (x_t, x_{t-k}) is estimated for all values of k in the vector lags.

REFERENCES

Abramson, I. S. (1982). On bandwidth variation in kernel estimates – a square root law. *Ann. Statist.*, **10**, 1217–23.

Aitchison, J. (1986). *The Statistical Analysis of Compositional Data.* Chapman & Hall, London.

Aitchison, J. and Aitken, C. G. G. (1976). Multivariate binary discrimination by the kernel method. *Biometrika*, **63**, 413–20.

Aitchison, J. and Lauder, I. J. (1985). Kernel dentisty estimation for compositional data. *Appl. Statist.*, **34**, 129–37.

Altman, N. S. (1990). Kernel smoothing of data with correlated errors. *J. Amer. Statist. Assoc.*, **85**, 749–59.

Anderson, A. H., Jensen, E. B. and Schou, G. (1981). Two-way analysis of variance with correlated errors. *Internat. Statist. Rev.*, **49**, 153–67.

Anderson, N. H. and Titterington, D. M. (1997). Some methods for investigating spatial clustering with epidemiological applications. *J. Roy. Statist. Soc. A*, **160**, to appear.

Anderson, N. H., Hall, P. and Titterington, D. M. (1994). Two-sample test statistics for measuring discrepancies between two multivariate probability density functions using kernel-based density estimates. *J. Multivariate Anal.*, **50**, 41–54.

Anderson, T. W. and Darling (1952). Asymptotic theory of certain "goodness of fit" criteria based on stochastic processes. *Ann. Math. Statist.*, **23**, 193–212.

Azzalini, A. (1981). A note on the estimation of a distribution function and quantiles by a kernel method. *Biometrika*, **68**, 326–8.

Azzalini, A. and Bowman, A. W. (1990). A look at some data on the Old Faithful geyser. *Appl. Statist.*, **39**, 357–65.

Azzalini, A. and Bowman, A. W. (1991). Nonparametric regression methods for repeated measurements. In: *Nonparametric Functional Estimation and Related Topics*, edited by G. G. Roussas. Kluwer Academic Publishers, Dordrecht, pp. 377–87.

Azzalini, A. and Bowman, A. W. (1993). On the use of nonparametric regression for checking linear relationships. *J. Roy. Statist. Soc. Ser. B*, **55**, 549–57.

Azzalini, A. and Chiogna, M. (1997). *S-Plus* tools for the analysis of repeated measures data. *Comput. Statist.*, **12**, 53–66.

Azzalini, A., Bowman, A. W. and Härdle, W. (1989). On the use of nonaparametric regression for model checking. *Biometrika*, **76**, 1–11.

Bailey, T. and Gatrell, A. (1995). *Interactive Spatial Data Analysis.* Longman Scientific and Technical, Harlow.

Beran, R. (1981). Nonparametric regression with randomly censored survival data. Technical report, University of California, Berkeley.

Bickel, P. J. and Rosenblatt, M. (1973). On some global measures of the deviations of density function estimates. *Ann. Statist.*, **1**, 1071–95.

Bissell, A. F. (1972). A negative binomial model with varying element sizes. *Biometrika*, **59**, 435–41.

Bithell, J. F. (1990). An application of density estimation to geographical epidemiology. *Statist. Med.*, **9**, 691–701.

Bjerve, S. and Doksum, K. (1993). Correlation curves - measures of association as functions of covariate values. *Ann. Statist.*, **21**, 890–902.

Block, E. (1952). Quantitative morphological investigations of the follicular system in women. Variations at different ages. *Acta Anat.*, **14**, 108–23.

Block, E. (1953). Quantitative morphological investigations of the follicular system in newborn female infants. *Acta Anat.*, **17**, 201–6.

Borchers, D. L., Buckland, S. T., Priede, I. G. and Ahmadi, S. (1997). Improving the daily egg production method using generalised additive models: estimation of the spawning stock biomass of mackerel (*Scomber scombrus*) and horse mackerel (*Trachurus trachurus*) in the NE Atlantic Ocean. *Canad. J. Fish. Aquat. Sci.*, to appear.

Bosq, D. (1996). *Nonparametric Statistics for Stochastic Processes.* Springer-Verlag, New York.

Bowman, A. W. (1984). An alternative method of cross-validation for the smoothing of density estimates. *Biometrika*, **711**, 353–60.

Bowman, A. W. (1992). Density-based tests for goodness of fit. *J. Statist. Comput. Simulation*, **40**, 1–13.

Bowman, A. W. and Foster, P. J. (1993). Adaptive smoothing and density-based tests of multivariate normality. *J. Amer. Statist. Assoc.*, **88**, 529–37.

Bowman, A. W. and Prvan, T. (1996). Cross-validation for the smoothing of distribution functions. Technical report, University of Glasgow.

Breiman, L. and Friedman, J. H. (1985). Estimating optimal transformations for multiple regression and correlation (with discussion). *J. Amer. Statist. Assoc.*, **80**, 580–619.

Breiman, L., Meisel, W. and Purcell, E. (1977). Variable kernel estimates of multivariate densities. *Technometrics*, **19**, 135–44.

Brillinger, D. R. (1980). Analysis of variance under time series models. In: *Handbook of Statistics*, **1**, *Analysis of Variance*, edited by P. R. Krishnaiah. North-Holland, Amsterdam.

Brillinger, D. R. (1981). *Time Series: Data Analysis and Theory* (expanded edition). J. Wiley & Sons, New York.

Brinkman, N. D. (1981). Ethanol fuel — a single-cylinder engine study of efficiency and exhaust emissions. *Soc. Automotive Eng. Trans.*, **90**, 1410–24.

Brockmann, M., Gasser, T. and Herrmann, E. (1993). Locally adaptive bandwidth choice for kernel regression estimators. *J. Amer. Statist. Assoc.*, **88**, 1302–9.

Buja, A., Donnell, D. and Stuetzle, W. (1986). Additive principal components. Technical report, University of Washington, Seattle.

Buja, A., Hastie, T. and Tibshirani, R. (1989). Linear smoothers and additive models (with discussion). *Ann. Statist.*, **17**, 453–555.

Čencov, N. N. (1962). Evaluation of an unknown distribution density from observations. *Soviet Math.*, **3**, 1559–62.

Cheng, B., Hall, P. and Titterington, D. M. (1997). On the shrinkage of local linear curve stimates. *Statistics and Computing*, **7**, 11–7.

Chu, C.-K. and Marron, J. S. (1991). Choosing a kernel regression estimator (with discussion). *Statist. Science*, **6**, 404–36.

Clark, R. M. (1977). Calibration, cross-validation and carbon-14. II. *J. Roy. Statist. Soc. Ser. A*, **143**, 177–94.

Cleveland, W. S. (1979). Robust locally weighted regression and smoothing scatterplots. *J. Amer. Statist. Assoc.*, **74**, 829–36.

Cleveland, W. S. and Devlin, S. J. (1988). Locally weighted regression: an approach to regression analysis by local fitting. *J. Amer. Statist. Assoc.*, **83**, 596–610.

Cleveland, W. S. and Loader, C. (1995). Smoothing by local regression: principles and methods (with discussion). *Comput. Statist.*, to appear.

Cleveland, W. S., Grosse, E. H. and Shyu, W. M. (1992). Local regression models. In: *Statistical Models in S*, edited by J. M. Chambers and T. Hastie. Wadsworth and Brooks/Cole, Pacific Grove, CA, pp. 309–76.

Cook, R. D. and Weisberg, S. (1994). Transforming a response variable for linearity. *Biometrika*, **81**, 731–7.

Copas, J. B. (1995). Local likelihood based on kernel censoring. *J. Roy. Statist. Soc. Ser. B*, **57**, 221–35.

Cox, D. D. (1986). Approximation theory of method of regularization estimators. In: *Contemporary Methematics* Vol. 59: *Function Estimates*, edited by J. S. Marron. American Mathematical Society, Providence, RI.

Cox, D. D. and Koh, E. (1989). A smoothing spline based test of model adequacy in polynomial regression. *Ann. Inst. Statist. Math.*, **41**, 383–400.

Cox, D. D., Koh, E., Wahba, G. and Yandell, B. S. (1988). Testing the (parametric) null model hypothesis in (semiparametric) partial and generalised spline models. *Ann. Statist.*, **16**, 113–19.

Davison, A. C. and Hinkley, D. V. (1997). *Bootstrap Methods and their Application.* Cambridge University Press, Cambridge.

Devroye, L. and Györfi, L. (1985). *Nonparametric Density Estimation: the L_1 View.* Wiley, New York.

Diblasi, A. and Bowman, A. W. (1997). Testing for constant variance in a linear model. *Statist. Probab. Lett.*, **33**, 95–103.

Diggle, P. J. (1985). A kernel method for smoothing point process data. *Appl. Statist.*, **34**, 138–47.

Diggle, P. J. (1990). A point process modelling approach to raised incidence of a rare phenomenon in the vicinity of a pre-specified point. *J. Roy. Statist.*

Soc. A, **153**, 349–62.

Diggle, P. J. and Al-Wasel, I. (1997). Spectral analysis of replicated biomedical time series (with discussion). *Appl. Statist.*, to appear.

Diggle, P. J. and Fisher, N. I. (1985). SPHERE: a contouring program for spherical data. *Comput. Geosciences*, **11**, 725–66.

Diggle, P. J. and Rowlingson, B. S. (1994). A conditional approach to point process modelling of raised incidence. *J. Roy. Statist. Soc. Ser. A*, **157**, 433–40.

Diggle, P. J., Gatrell, A. C. and Lovett, A. A. (1990). Modelling the prevalence of cancer of the larynx in part of Lancashire: a new methodology for spatial epidemiology. In: *Spatial Epidemiology*, edited by R. W. Thomas, London Papers in Regional Science, **21**. Pion, London.

Dobson, A. J. (1990). *An Introduction to Generalized Linear Models* (2nd edn). Chapman & Hall, London.

Doksum, K. A. and Yandell, B. S. (1982). Properties of regression estimates based on censored survival data. In: *A Festschrift for Erich Lehmann*, edited by P. J. Bickel, K. A. Doksum and J. L. Hodges, Jr. Wadsworth International Group, Belmont, CA.

Doksum, K., Blyth, S., Bradlow, E., Meng, X.-L. and Zhao, H. (1994). Correlation curves as local measures of variance explained by regression. *J. Amer. Statist. Assoc.*, **89**, 571–82.

Draper, N. R. and Guttman, I (1980). Incorporating overlap effects from neighbouring units into response surface models. *Appl. Statist.*, **29**, 128–34.

Duan, N. and Li, K. C. (1991). Slicing regression: a link-free regression method. *Ann. Statist.*, **19**, 505–30.

Dugmore, A. J., Larsen, G., Newton, A. J. and Sugden, D. E. (1992). Geochemical stability of finegrained silicic tephra in Iceland and Scotland. *J. Quatern. Sci.*, **7**, 173–83.

Epanechnikov, V. A. (1969). Non-parametric estimation of a multivariate probability density. *Theory Probab. Appl.*, **14**, 153–8.

Eubank, R. L. (1988). *Spline Smoothing and Nonparametric Regression.* Marcel Dekker, New York and Basel.

Eubank, R. L. and Speckman, P. L. (1993). Confidence bands in nonparametric regression. *J. Amer. Statist. Assoc.*, **88**, 1287–1301.

Eubank, R. L. and Spiegelman, C. H. (1990). Testing the goodness-of-fit of a linear model via nonparametric regression techniques. *J. Amer. Statist. Assoc.*, **85**, 387–92.

Faddy, M. J. and Gosden, R. G. (1996). A model conforming the decline in follicle numbers to the age of menopause in women. *Human Reproduction*, **11**, 1484–6.

Faddy, M. J. and Jones, M. C. (1997). Modelling and analysis of data that exhibit temporal decay using local quadratic smoothing. Technical report.

Falk, M. (1983). Relative efficiency and deficiency of kernel type estimators of smooth distribution functions. *Statist. Neerlandica*, **37**, 73–83.

Fan, J. (1992). Design-adaptive nonparametric regression. *J. Amer. Statist. Assoc.*, **87**, 998–1004.

Fan, J. (1993). Local linear regression smoothers and their minimax efficiencies. *Ann. Statist.*, **21**, 196–216.

Fan, J. and Gijbels, I. (1992). Variable bandwidth and local linear regression smoothers. *Ann. Statist.*, **20**, 2008–36.

Fan, J. and Gijbels, I. (1994). Censored regression: local linear approximations and their applications. *J. Amer. Statist. Assoc.*, **89**, 560–70.

Fan, J. and Gijbels, I. (1995). Data-driven bandwidth selection in local polynomial fitting: variable bandwidth and spatial adaptation. *J. Roy. Statist. Soc. Ser. B*, **57**, 371–94.

Fan, J. and Gijbels, I. (1996). *Local polynomial modelling and its applications.* Chapman & Hall, London.

Fan, J. and Marron, J. S. (1994). Fast implementations of nonparametric curve estimators. *J. Comput. Graph. Statist.*, **3**, 35–56.

Fan, J., Heckmann, N. E. and Wand, M. P. (1995). Local polynomial kernel regression for generalized linear models and quasi likelihood functions. *J. Amer. Statist. Assoc*, **90**, 141–50.

Fernholz, L. T. (1991). Almost sure convergence of smoothed empirical distribution functions. *Scand. J. Statist.*, **18**, 255–62.

Filliben, J. J. (1975). The probability plot correlation coefficient test for normality. *Technometrics*, **17**, 111–17.

Firth, D., Glosup, J. and Hinkley, D. V. (1991). Model checking with nonparametric curves. *Biometrika*, **78**, 245–52.

Fisher, N. I., Lewis, T. and Embleton, B. J. J. (1987). *Statistical Analysis of Spherical Data.* Cambridge University Press, Cambridge.

Fowlkes, E. B. (1987). Some diagnostics for binary logistic regression via smoothing. *Biometrika*, **74**, 503–15.

Friedman, J. H. (1991). Multivariate adaptive regression splines (with discussion). *Ann. Statist.*, **19**, 1–141.

Friedman, J. H. and Stuetzle, W. (1981). Projection pusuit regression. *J. Amer. Statist. Assoc.*, **76**, 817–23.

Gasser, T. and Müller, H.-G. (1979). Kernel estimation of regression functions. In *Smoothing Techniques for Curve Estimation*, edited by T. Gasser and M. Rosenblatt. Springer-Verlag, Heidelberg, pp. 23–68.

Gasser, T., Müller, H.-G., Köhler, W., Molinari, L. and Prader, A. (1984). Nonparametric analysis of growth curves. *Ann. Statist.*, **12**, 210–29. Correction: **12, 1588.**

Gasser, T., Sroka, L. and Jennen-Steinmetz, C. (1986). Residual variance and residual pattern in nonlinear regression. *Biometrika*, **73**, 625–33.

Gasser, T., Kneip, A. and Köhler, W. (1991). A flexible and fast method for automatic smoothing. *J. Amer. Statist. Assoc.*, **86**, 643–52.

Gentleman, R. and Crowley, J. (1991). Graphical methods for censored data. *J. Amer. Statist. Assoc.*, **86**, 678–83.

Granovsky, B. L. and Müller, H.-G. (1991). Optimizing kernel methods: a unifying variational principle. *Internat. Statist. Rev.*, **59**, 373–88.

Green, P. J. (1987). Penalized likelihood for general semi-parametric regression models. *Internat. Statist. Rev.*, **55**, 245–59.

Green, P. J. and Silverman, B. W. (1994). *Nonparametric Regression and Generalized Linear Models: A Roughness Penalty Approach.* Chapman & Hall, London.

Green, P. J., Jennison, C. and Seheult, A. (1985). Analysis of field experiments by least squares smoothing. *J. Roy. Statist. Soc. Ser. B*, **47**, 299–315.

Grizzle, J. E. and Allen, D. M. (1969). Analysis of growth and dose response curves. *Biometrics*, **25**, 357–81.

Györfi, L., Härdle, W., Sarda, P. and Vieu, P. (1989). *Nonparametric Curve Estimation from Time Series.* Springer-Verlag, Berlin.

Hall, P. (1990). On the bias of variable bandwidth curve estimators. *Biometrika*, **77**, 529–35.

Hall, P. (1992). Effect of bias estimation on coverage accuracy of bootstrap confidence intervals for a probability density. *Ann. Statist.*, **20**, 675–94.

Hall, P. and Hart, J. D. (1990). Bootstrap test for difference between means in nonparametric regression. *J. Amer. Statist. Assoc.*, **85**, 1039–49.

Hall, P. and Johnstone, I. M. (1992). Empirical functionals and efficient smoothing parameter selection (with discussion). *J. Roy. Statist. Soc. Ser. B*, **54**, 475–530.

Hall, P. and Marron, J. S. (1988). Variable window width kernel estimates of probabilioty densities. *Probab. Theory Related Fields*, **80**, 37–49.

Hall, P. and Marron, J. S. (1990). On variance estimation in nonparametric regression. *Biometrika*, **77**, 415–19.

Hall, P. and Titterington, D. M. (1988). On confidence bands in nonparametric density estimation and regression. *J. Multivariate Anal.*, **27**, 228–54.

Hall, P., Kay, J. W. and Titterington, D. M. (1990). Asymptotically optimal difference-based estimation of variance in nonparametric regression. *Biometrika*, **77**, 521–8.

Hall, P., Marron, J. S. and Park, B. U. (1992). Smoothed cross-validation. *Probab. Theory Related Fields*, **92**, 1–20.

Hand, D. J., Daly, F., Lunn, A. D., McConway, K. J. and Ostrowsky, E. (1994). *A Handbook of Small Data Sets.* Chapman & Hall, London.

Härdle, W. (1989). Asymptotic maximal deviation of M-smoothers. *J. Multivariate Anal.*, **29**, 163–79.

Härdle, W. (1990). *Applied Nonparametric Regression.* Cambridge University Press, London.

Härdle, W. and Bowman, A. W. (1988). Bootstrapping in nonparametric regression: local adaptive smoothing and confidence bands. *J. Amer. Statist. Assoc.*, **83**, 102–10.

Härdle, W. and Marron, J. S. (1991). Bootstrap simultaneous error bars for nonparametric regression. *Ann. Statist.*, **19**, 778–96.

Härdle, W. and Tsybakov, A. B. (1997). Local polynomial estimators of the volatility functions in nonparametric autoregression. *J. Econometrics*, to appear.

Härdle, W. and Vieu, P. (1992). Kernel regression smoothing of time series. *J. Time Ser. Anal.*, **13**, 209–32.

Härdle, W., Hall, P. and Marron, J. S. (1988). How far are automatically chosen smoothing parameters from their optimum? (with discussion). *J. Amer. Statist. Assoc.*, **83**, 86–101.

Härdle, W., Lütkepohl, H. and Chen, R. (1997). A review of nonparametric time series analyis. *Intern. Statist. Rev.*, **65**, 49–72.

Hart, J. D. (1991). Kernel regression estimation with time series errors. *J. Roy. Statist. Soc. Ser. B*, **53**, 173–87.

Hart, J. D. and Wehrly, T. E. (1986). Kernel regression estimation using repeated measurements data. *J. Amer. Statist. Assoc.*, **81**, 1080–8.

Hastie, T. and Loader, C. (1993). Local regression: automatic kernel carpentry (with discussion). *Statist. Sc.*, **8**, 120–43.

Hastie, T. and Tibshirani, R. (1990). *Generalized Additive Models.* Chapman & Hall, London.

Hastie, T. and Tibshirani, R. (1993). Varying-coefficient models. *J. Roy. Statist. Soc. Ser. B*, **55**, 757-96.

Herrmann, E., Gasser, T. and Kneip, A. (1992). Choice of bandwidth for kernel regression when residuals are correlated. *Biometrika*, **79**, 783–95.

Hjört, N. L. and Jones, M. C. (1996). Locally parametric nonparametric density estimation. *Ann. Statist.*, **24**, 1619–47.

Hogg, R. V. (1979). Statistical robustness: one view of its use in applications today. *Amer. Statist.*, **33**, 108–16.

Hosmer, D. W.Jr. and Lemeshow, S. (1989). *Applied Logistic Regression.* Wiley, New York.

Jane's (1978). *Jane's Encyclopaedia of Aviation.* Jane's, London.

Joe, H. (1989). Relative entropy measures of multivariate dependence. *J. Amer. Statist. Assoc.*, **84**, 157–64.

Johnson, N. L. and Kotz, S. (1972). *Distributions in Statistics: Continuous Univariate Distributions, Vol. II.* Wiley, New York.

Jones, M. C. (1996). The local dependence function. *Biometrika*, **83**, 899–904.

Jones, M. C., Davies, S. J. and Park, B. U. (1994). Versions of kernel-type regression estimators. *J. Amer. Statist. Assoc.*, **89**, 825–32.

Jones, M. C., Marron, J. S. and Sheather, S. J. (1996). A brief survey of bandwidth selection for density estimation. *J. Amer. Statist. Assoc.*, **91**, 401–7.

Kelsall, J. E. and Diggle, P. J. (1995a) Kernel estimation of relative risk. *Bernoulli*, **1**, 3–16.

Kelsall, J. E. and Diggle, P. J. (1995b) Non-parametric estimation of spatial variation in relative risk. *Statist. Med.*, **14**, 2335–42.

Kent, J. T. (1983). Information gain and a general measure of correlation. *Biometrika*, **70**, 163–73.

King, E. C., Hart, J. D. and Wehrly, T. E. (1991). Testing the equality of two regression curves using linear smoothers. *Statist. Probab. Lett.*, **12**, 239–47.

Knafl, G., Sacks, J. and Ylvisaker, D. (1985). Confidence bands for regression functions. *J. Amer. Statist. Assoc.*, **80**, 683–91.

Kronmal, R. A. and Tarter, M. E. (1968). The estimation of probability densities and cumulatives by Fourier series methods. *J. Amer. Statist. Assoc.*, **63**, 925–52.

Landwehr, J. M., Pregibon, D. and Shoemaker, A. C. (1984). Graphical methods for assessing logistic regression models (with discussion). *J. Amer. Statist. Assoc.*, **79**, 61–83.

le Cessie, S. and van Houwelingen, J. C. (1991). A goodness-of-fit test for binary regression models, based on smoothing methods. *Biometrics*, **47**, 1267–82.

le Cessie, S. and van Houwelingen, J. C. (1993). Building logistic models by means of a nonparametric goodness of fit test: a case study. *Statist. Neerlandica*, **47**, 97–110.

le Cessie, S. and van Houwelingen, J. C. (1995). Testing the fit of a regression model via score tests in random effects models. *Biometrics*, **51**, 600–14.

Li, K. C. (1991). Sliced inverse regression for dimension reduction (with discussion). *J. Amer. Statist. Assoc.*, **86**, 316–42.

Loftsgaarden, D. O. and Quesenberry, C. P. (1965). A nonparametric estimate of a multivariate density function. *Ann. Math. Statist.*, **36**, 1049–51.

Marron, J. S. (1993). Discussion of 'Practical performance of several data-driven bandwidth selectors' by Park and Turlach. *Comput. Statist.*, **8**, 17–19.

Marron, J. S. and Wand, M. P. (1992). Exact mean integrated squared error. *Ann. Statist.*, **20**, 712–36.

Moore, D. S. and Yackel, J. W. (1977). Consistency properties of nearest neighbor density function estimators. *Ann. Statist.*, **5**, 143–54.

Müller, H.-G. (1988). *Nonparametric Regression Analysis of Longitudinal Data.* Lecture Notes in Statist., **46**. Springer-Verlag, Berlin.

Nadaraya, E. A. (1964a). Some new estimates for distribution functions. *Theory Probab. Appl.*, **9**, 497–500.

Nadaraya, E. A. (1964b) On estimating regression. *Theory Probab. Appl.*, **10**, 186–90.

Nychka, D. (1988). Bayesian confidence intervals for a smoothing spline. *J. Amer. Statist. Assoc.*, **83**, 1134–43.

Nychka, D. (1990). The average posterior variance of a smoothing spline and a consistent estimate of the average squared error. *Ann. Statist.*, **18**, 415–28.

Park, B. U. and Marron, J. S. (1990). Comparison of data-driven bandwidth selectors. *J. Amer. Statist. Assoc.*, **85**, 66–72.

Parzen, E. (1962). On the estimation of a probability density and mode. *Ann. Math. Statist.*, **33**, 1065–76.

Pearson, G. W. and Qua, F. (1993). High precision ^{14}C measurement of Irish oaks to show the natural ^{14}C variations from AD 1840–5000 BC: a correction. *Radiocarbon*, **35**, 105–23.

Poiner, I. R., Blaber, S. J. M., Brewer, D. T., Burridge, C. Y., Caesar, D., Connell, M., Dennis, D., Dews, G. D., Ellis, A. N., Farmer, M., Fry, G. J., Glaister, J., Gribble, N., Hill, B. J., Long, B. G., Milton, D. A., Pitcher, C. R., Proh, D., Salini, J. P., Thomas, M. R., Toscas, P., Veronise, S., Wang, Y. G., Wassenberg, T. J. (1997). The effects of prawn trawling in the far northern section of the Great Barrier Reef. Final report to GBRMPA and FRDC on 1991–96 research. CSIRO Division of Marine Research, Queensland Dept. of Primary Industries.

Priede, I. G. and Watson, J. J. (1993). An evaluation of the daily egg production method for estimating biomass of Atlantic mackerel (*Scomber scombrus*). *Bull. Marine Sc.*, **53**, 891–911.

Priede, I. G., Raid, T. and Watson, J. J. (1995). Deep-water spawning of Atlantic mackerel *Scomber scombrus*, west of Ireland. *J. Marine Biol. Assoc.*, **75**, 849–55.

Priestley, M. B. (1981). *Spectral Analysis and Time Series* (2 vol.). Academic Press, London.

Ratkowsky, D. A. (1983). *Nonlinear Regression Modelling*. Dekker, New York.

Raz, J. (1989). Analysis of repeated measurements using nonparametric smoothers and randomization tests. *Biometrics*, **45**, 851–71.

Raz, J. (1990). Testing for no effect when estimating a smooth function by nonparametric regression: a randomization approach. *J. Amer. Statist. Assoc.*, **85**, 132–38.

Rice, J. (1984). Bandwidth choice for nonparametric regression. *Ann. Statist.*, **12**, 1215–30.

Rice, J. A. and Silverman, B. W. (1991). Estimating the mean and the covariance structure nonparametrically when the data are curves. *J. Roy. Statist. Soc. Ser. B*, **53**, 233–243.

Richardson, S. J., Senikas, V. and Nelson, J. F. (1987). Follicular depletion during the menopausal transition: evidence for accelerated loss and ultimate exhaustion. *J. Clinical Endocrinology Metabolism*, **65**, 1231–7.

Ripley, B. D. (1994). Neural networks and related methods for classification (with discussion). *J. Roy. Statist. Soc. Ser. B*, **56**, 409–56.

Robinson, P. M. (1983). Nonparametric estimators for time series. *J. Time Ser. Anal.* **4**, 185–207.

Rosenberg, P. S. (1995). Hazard function estimation using B-splines. *Biometrics*, **51**, 874–87.

Rosenblatt, M. (1956). Remarks on some non-parametric estimates of a density function. *Ann. Math. Statist.*, **27**, 832–7.

Rosenblatt, M. (1971). Curve estimates. *Ann. Math. Statist.*, **42**, 1815–42.

Roussas, G. G., editor (1991). *Nonparametric Functional Estimation and Related Topics*. Kluwer Academic Publishers, Dordrecht.

Royston, P. and Altman, D. G. (1994). Regression using fractional polynomials of continuous covariates: parsimonious parametric modelling (with discussion). *Appl. Statist.*, **43**, 429–67.

Rudemo, M. (1982). Empirical choice of histograms and kernel density estimators. *Scand. J. Statist.*, **9**, 65–78.

Ruppert, D. and Wand, M. P. (1994). Multivariate locally weighted least squares regression. *Ann. Statist.*, **22**, 1346–70.

Ruppert, D., Sheather, S. J. and Wand, M. P. (1995). An effective bandwidth selector for local least squares regression. *J. Amer. Statist. Assoc.*, **90**, 1257–70.

Ryan, B. F., Joiner, B. L. and Ryan. T. A., Jr.(1985). *Minitab Handbook* (2nd edn). PWS-Kent Publishing Company, Boston.

Saviotti, P. P. and Bowman, A. B. (1984). Indicators of output of technology. In: *Proceedings of the ICSSR/SSRC Workshop on Science and Technology in the 1980's*, edited by M. Gibbons *et al.*. Harvester Press, Brighton.

Schwartz, S. C. (1967). Estimation of probability density by an orthogonal series. *Ann. Math. Statist.*, **38**, 1261–5.

Scott, D. W. (1992). *Multivariate Density Estimation: Theory, Practice and Visualisation.* Wiley, New York.

Scott, D. W. and Terrell, G. (1987). Biased and unbiased cross-validation in density estimation. *J. Amer. Statist. Assoc.*, **82**, 1131–46.

Scott, D. W., Tapia, R. A. and Thompson, J. R. (1977). Kernel density estimation revisted. *Nonlinear Anal.*, **1**, 339-72.

Scott, E. M., Baxter, M. S. and Aitchison, T. C. (1984). A comparison of the treatment of errors in radiocarbon dating calibration methods. *J. Archaeol. Science*, **11**, 455–66.

Seifert, B. and Gasser, T. and Wolf, A. (1993). Nonparametric estimation of residual variance revisited. *Biometrika*, **80**, 373–84.

Shapiro, S. S. and Wilk, M. B. (1965). An analysis of variance test for normality (complete samples). *Biometrika*, **52**, 591–611.

Sheather, S. J. and Jones, M. C. (1991). A reliable data-based bandwidth selection method for kernel density estimation. *J. Roy. Statist. Soc. Ser. B*, **53**, 683–90.

Shi, M., Weiss, R. E. and Taylor, J. M. G. (1996). An analysis of pediatric AIDS CD4 counts using flexible random curves. *Appl. Statist.*, **45**, 151–63.

Silverman, B. W. (1981). Using kernel density estimates to investigate multimodality. *J. Roy. Statist. Soc. Ser. B*, **43**, 97–9.

Silverman, B. W. (1986). *Density Estimation for Statistics and Data Analysis.* Chapman & Hall, London.

Simonoff, J. S. (1996). *Smoothing Methods in Statistics.* Springer-Verlag, New York.

Speckman, P. L. (1988). Kernel smoothing in partial linear models. *J. Roy. Statist. Soc. Ser. B*, **50**, 413–36.

Staniswalis, J. G. and Severini, T. A. (1991). Diagnostics for assessing regression models. *J. Amer. Statist. Assoc.*, **86**, 684–92.

Stephens, M. A. (1974). EDF statistics for goodness of fit and some comparisons. *J. Amer. Statist. Assoc.*, **69**, 730–37.

Stone, C. J. (1984). An asymptotically optimal window selection rule for kernel density estimates. *Ann. Statist.*, **12**, 1285–97.

Stone, M. A. (1974). Cross-validatory choice and assessment of statistical predictions. *J. Roy. Statist. Soc. Ser. B*, **36**, 111–47.

Taylor, C. C. (1989). Bootstrap choice of the smoothing parameter in kernel density estimation. *Biometrika*, **76**, 705–12.

Terrell, G. R. (1990). The maximal smoothing principle in density estimation. *J. Amer. Statist. Assoc.*, **85**, 470–7.

Terrell, G. R. and Scott, D. W. (1985). Oversmoothed nonparametric density estimates. *J. Amer. Statist. Assoc.*, **80**, 209–14.

Tibshirani, R. (1988). Estimating optimal transformations for regression via additivity and variance stabilization. *J. Amer. Statist. Assoc.*, **83**, 394-405.

Tibshirani, R. and Hastie, T. (1987). Local likelihood estimation. *J. Amer. Statist. Assoc.*, **82**, 559–67.

Tukey, J. W. (1977). *Exploratory Data Analysis.* Addison-Wesley, Reading, MA.

Vasicek, D. (1976). A test for normality based on sample entropy. *J. Roy. Statist. Soc. Ser. B*, **38**, 54–9.

Wahba, G. (1983). Bayesian 'confidence intervals' for the cross-validated smoothing spline. *J. Roy. Statist. Soc. Ser. B*, **45**, 133–50.

Wahba, G. (1990). *Spline Models for Observational Data.* SIAM, Philadelphia.

Wand, M. P. and Jones, M. C. (1995). *Kernel Smoothing.* Chapman & Hall, London.

Wand, M. P., Marron, J. S. and Ruppert, D. (1991). Transformations in density estimation (with comments). *J. Amer. Statist. Assoc.*, **86**, 342–61.

Watson, G. S. (1964). Smooth regression analysis. *Sankhyā, Ser. A*, **26**, 359–72.

Watson, G. S. and Leadbetter, M. R. (1964). Hazard analysis I. *Biometrika*, **51**, 175–84.

Watson, J. J., Priede, I. G., Witthames, P. R. and Owari-Wadunde, A. (1992). *J. Fish Biol.*, **40**, 591–8.

Weidong, P., Xianmin, Z., Xiaomin, C., Crompton, D. W. T., Whitehead, R. R., Jiangqin, X., Haigeng, W., Jiyuan, P., Yang, Y., Weixing, W., Kaiwu, X. and Yongxing, Y. (1996). *Ascaris*, people and pigs in a rural community of Jiangxi province, China. *Parasitology*, **113**, 545–57.

Weisberg, S. (1985). *Applied Linear Regression* (2nd edn). Wiley, New York.

Weisberg, S. and Welsh, A. H. (1994). Adapting for the missing link. *Ann. Statist.*, **22**, 1674–1700

Whittle, P. (1958). On the smoothing of probability density functions. *J. Roy. Statist. Soc. Ser. B*, **55**, 549–57.

Wright, E. M. and Bowman, A. W. (1997). Exploration of survival data using nonparametric quantile estimators. Technical report, University of Glasgow.

Young, S. G. (1996). Graphics and Inference in Nonparametric Modelling. Ph.D. thesis, University of Glasgow.

Young, S. G. and Bowman, A. W. (1995). Non-parametric analysis of covariance. *Biometrics*, **51**, 920–31.

Zeger, S. L. and Diggle, P. J. (1994). Semi-parametric models for longitudinal data with application to CD4 cell number in HIV seroconverters. *Biometrics*, **50**, 689–99.

AUTHOR INDEX

INDEX

Lightning Source UK Ltd.
Milton Keynes UK
UKOW01n1423150817
307361UK00001B/26/P